AF839796

A Novel Approach to Seawater Desalination Using Dual-Staged Nanofiltration

Awwa Research Foundation

About the Awwa Research Foundation

The Awwa Research Foundation (AwwaRF) is a member-supported, international, nonprofit organization that sponsors research to enable water utilities, public health agencies, and other professionals to provide safe and affordable drinking water to consumers.

The Foundation's mission is to advance the science of water to improve the quality of life. To achieve this mission, the Foundation sponsors studies on all aspects of drinking water, including supply and resources, treatment, monitoring and analysis, distribution, management, and health effects. Funding for research is provided primarily by subscription payments from approximately 1,000 utilities, consulting firms, and manufacturers in North America and abroad. Additional funding comes from collaborative partnerships with other national and international organizations, allowing for resources to be leveraged, expertise to be shared, and broad-based knowledge to be developed and disseminated. Government funding serves as a third source of research dollars.

From its headquarters in Denver, Colorado, the Foundation's staff directs and supports the efforts of more than 800 volunteers who serve on the board of trustees and various committees. These volunteers represent many facets of the water industry, and contribute their expertise to select and monitor research studies that benefit the entire drinking water community.

The results of research are disseminated through a number of channels, including reports, the Web site, conferences, and periodicals.

For subscribers, the Foundation serves as a cooperative program in which water suppliers unite to pool their resources. By applying Foundation research findings, these water suppliers can save substantial costs and stay on the leading edge of drinking water science and technology. Since its inception, AwwaRF has supplied the water community with more than $300 million in applied research.

More information about the Foundation and how to become a subscriber is available on the Web at **www.awwarf.org.**

A Novel Approach to Seawater Desalination Using Dual-Staged Nanofiltration

Prepared by:
Yann A. Le Gouellec and **David A. Cornwell**
EE&T, Inc., Newport News, VA 23606

Robert C. Cheng, **Tai J. Tseng**, **Diem X. Vuong**, and **Kevin L. Wattier**
Long Beach Water Department, Long Beach, CA 90806

and

Catherine J. Harrison and **Amy E. Childress**
Department of Civil Engineering, University of Nevada, Reno, NV 89557

Jointly sponsored by:
Awwa Research Foundation
6666 West Quincy Avenue, Denver, CO 80235-3098

and

Long Beach Water Department
Long Beach, CA 90807

Published by:

Awwa Research Foundation

American Water Works Association

IWA Publishing

DISCLAIMER

This study was jointly funded by the Awwa Research Foundation (AwwaRF) and the Long Beach Water Department (LBWD). AwwaRF and LBWD assume no responsibility for the content of the research study reported in this publication or for the opinions and statements of fact expressed in the report. The mention of trade names for commercial products does not represent or imply the approval or endorsement of AwwaRF or LBWD. This report is presented solely for informational purposes.

Printed on recycled paper

CONTENTS

TABLES

FIGURES

FOREWORD

The Awwa Research Foundation is a nonprofit corporation that is dedicated to the implementation of a research effort to help utilities respond to regulatory requirements and traditional high-priority concerns of the industry. The research agenda is developed through a process of consultation with subscribers and dinking water professionals. Under the umbrella of a Strategic Research Plan, the Research Advisory Council prioritizes the suggested projects based upon current and future needs, applicability, and past work; the recommendations are forwarded to the Board of Trustees for final selection. The foundation also sponsors research projects through the unsolicited proposal process; the Collaborative Research, Research Applications, and Tailored Collaboration programs; and various joint research efforts with organizations such as the U.S. Environmental Protection Agency, the U.S. Bureau of Reclamation, and the Association of California Water Agencies.

This publication is a result of one of these sponsored studies, and it is hoped that its findings will be applied in communities throughout the world. The following report serves not only as a means of communication the results of the water industry's centralized research program but also as a tool to enlist the further support of the nonmember utilities and individuals.

Projects are managed closely from their inception to the final report by the foundation's staff and large cadre of volunteers who willingly contribute their time and expertise. The foundation serves a planning and management function and awards contracts to other institutions such as water utilities, universities, and engineering firms. The funding for this research effort comes primarily from the Subscription Program, through which water utilities subscribe to the research program and make an annual payment proportionate to the volume of water they deliver and consultants and manufacturers subscribe based on their annual billings. The program offers a cost-effective and fair method for funding research in the public interest.

A broad spectrum of water supply issues is addressed by the foundation's research agenda: resources, treatment and operations, distribution and storage, water quality and analysis, toxicology, economics, and management. The ultimate purpose of the coordinated effort is to assist water suppliers to provide the highest possible quality of water economically and reliably. The true benefits are realized when the results are implemented at the utility level. The foundation's trustees are pleased to offer this publication as a contribution toward that end.

Walter J. Bishop
Chair, Board of Trustees
Awwa Research Foundation

Robert C. Renner, P.E.
Executive Director
Awwa Research Foundation

ACKNOWLEDGMENTS

The authors gratefully acknowledge financial support from Metropolitan Water District of Southern California (MWD) and thank the following Long Beach Water Department (LBWD) individuals who worked many hours to provide pilot data to the researchers:

Cynthia E. Andrews-Tate, Laboratory Manager
Janel Grebel and Phu Nguyen, Engineering Assistants

In addition, Richard A. Brown from EE&T, Inc. was responsible for the modeling effort on blending strategies. The advice of the Project Advisory Committee (PAC) – including John Balliew, El Paso Water Utility; Steven J. Duranceau, Boyle Engineering Corporation; Christine A. Owen, Tampa Bay Water – and the help of AwwaRF project manager, Jason Allen, are appreciated.

EXECUTIVE SUMMARY

INTRODUCTION

As the quantity and quality of inland water sources decline, more coastal municipalities are looking at seawater desalination as a potential source of drinking water. While reverse osmosis (RO) is a proven technology for seawater desalination, the energy required is high as compared to the energy required for treatment of other source waters (e.g., surface or ground water). An alternative to desalting seawater by the traditional seawater RO (SWRO) method is a novel dual-staged nanofiltration (NF^2) process developed by the Long Beach Water Department (LBWD). Although NF membranes have been primarily used for hardness and disinfection by-product (DBP) precursor removal, this novel NF system is able to desalt seawater by treating the first-stage permeate through a second stage in order to produce finished water with salinity levels that meet drinking water standards. In addition to seawater desalting, the NF^2 process has additional benefits compared to single-pass SWRO, namely:

- Two physical barriers:
  - Provide system redundancy for increased water quality protection
  - Allow boron removal through pH increase in the second stage
- Lower pressure requirements resulting in energy savings:
  - The comparative "looseness" of NF membranes means that the membrane hydraulic resistance is less than for SWRO

The capability of dual-staged nanofiltration for seawater desalination to reduce operational costs was assessed through a research plan divided into four phases:

- Theoretical basis of the system (governing equations, membrane suitability, etc.)
- Operational optimization approaches (pilot tests and use of a predictive model)
- Preliminary strategies for blending desalinated water with existing finished water
- Verification of inherent redundancy of the process with viral challenge tests

THEORETICAL BASIS

Membrane Characterization

Three commercially-available NF membranes were selected for this study based upon their designation as NF membranes by the manufacturers and their salt rejection characteristics (FilmTec NF-90, TriSep TS-80, and Saehan NE-90). The zeta potential (surface charge), contact angle (hydrophilicity), and surface roughness of each membrane were evaluated to possibly relate membrane surface properties to a selection protocol.

All membranes were negatively charged in the pH range tested and exhibited an increasingly negative zeta potential as pH increased. Negatively-charged membranes tend to have higher level of rejection for anions but can also attract cations that could form a bridge between the membrane surface and negatively-charged contaminants. Contact angle values between the three membranes were close to each other and showed them to be overall hydrophilic (contact angle ranged 49° to 55°), which indicate a reduced tendency to foul. Water convection and particle transport occur preferentially in the valleys of rough membranes where the section is thinnest. A rough membrane surface would therefore experience substantial fouling because more particles would become trapped in the valleys. Based on roughness measurements, the NE-90 membrane appears to have a lesser tendency to foul than the other two membranes.

Governing Equations

The performance-predicting model equations were based on the solution-diffusion model and developed such that permeate flow rate and water quality could be determined based on measurable parameters (temperature, pressure, membrane area, feed concentration and flow) and using membrane-specific salt and water permeability coefficients derived from water flux and rejection results obtained in benchscale performance testing. The predictive model uses the following four equations in an iterative fashion to each membrane element of both stages:

$$\pi = R_g\, T \sum_i n_i \frac{C_{f_i}}{M_i} \tag{ES.1}$$

$$\frac{C_{b_i}}{C_{f_i}} = \frac{k_w\left(P - \sigma R_g T \sum_i n_i \frac{C_{f_i}}{M_i}\right)\left(\frac{Q_f}{A} - k_i\right) + \frac{Q_f\, k_i}{A}}{\left(\frac{Q_f}{A} - k_w\left(P - \sigma R_g T \sum_i n_i \frac{C_{f_i}}{M_i}\right)\right)\left(k_w\left(P - \sigma R_g T \sum_i n_i \frac{C_{f_i}}{M_i}\right) + k_i\right)} \tag{ES.2}$$

$$\frac{C_{p_i}}{C_{f_i}} = \frac{k_i}{k_w\left(P - \sigma R_g\, T \sum_i n_i \frac{C_{f_i}}{M_i}\right) + k_i} \tag{ES.3}$$

$$Q_p = A k_w\left(P - \sigma R_g\, T \sum_i n_i \frac{C_{f_i}}{M_i}\right) \tag{ES.4}$$

where π is the osmotic pressure (psi), R_g is the gas constant (1.206 psi.L / mol.°K), T is the temperature (°K), n is the number of dissociating ions, M is the molecular weight (g/mol), C is the concentration (g/L), P is the applied pressure (psi), Q is the flow (gpd), A is the membrane area (ft^2), σ is the membrane reflection coefficient (an indication of the selectivity of the membrane, $0< \sigma <1$), and k is the permeability coefficient (gfd/psi for k_w and gfd for k_i). Subscripts f, p, b, i, and w refer to feed, permeate, brine, salt i, and water, respectively.

Membrane Performance

A systematic investigation of the water flux and salt rejection of the NF membranes was performed using stirred-cell and bench-scale membrane test units. These tests were used with the membrane surface characteristics in the membrane selection protocol. They also provide the information necessary (i.e., the water and salt permeability coefficients) to model process performance at the pilot scale for the operation optimization phase. It was important to obtain individual salt mass transfer coefficients for each ion because NF membranes reject ions of various valence and size differently, as compared to SWRO membranes that provide uniformly high rejection for most ions. Feed concentrations were selected in order to bracket the expected salt concentration range in the feed of Stage 1 and Stage 2 for the pilot plant. Operating pressures and temperatures were similar to those tested in the pilot plant tests.

Several general trends were observed from the data. Values for k_w decreased with increasing salt concentration. This behavior suggests a reversible effect of the feed salinity (through concentration polarization) on the membrane surface. Because it is generally accepted that such a coefficient is constant for RO membranes, this could mean that the surface of NF membranes is more sensitive to changes in feed salinity (possibly because NF membranes are not as tight as RO membranes). Values of k_i increased with increasing temperature, and the k_i for monovalent ions is higher than for divalent ions, which is expected for NF membranes.

Table ES.1 shows k_w and k_i for the synthetic seawater experiments that simulated Stage 1 of the pilot plant. The NF-90 and NE-90 membranes had similar k_w values of approximately 0.1 gfd/psi. This k_w value matched well with the k_w observed for the single salt NaCl experiment of similar ionic strength to the seawater. Both membranes also had k_i values for divalent ions that were significantly less than for monovalent ions. The NF-90 membrane had lower k_i values than NE-90 for all ions. Table ES.2 shows k_i and k_w for the simulated Stage 2 experiments, which used Stage 1 permeate obtained with two different pressures (500 and 575 psi). Stage 2 k_w values are higher than for Stage 1 because feed concentration and the concentration polarization (CP) layer are smaller. The k_w value for each membrane was constant the whether feed was Stage 1 permeate obtained at 575 psi or 500 psi. The k_w for NF-90 exhibited a larger increase than NE-90 from Stage 1 to Stage 2 due to its higher ion rejection. The k_i values for all ions were lower in Stage 2 than Stage 1, again due to lower feed concentrations. Similar to Stage 1 results, the k_i values for divalent ions were substantially lower than for monovalent ions.

Critical Flux

The purpose of this study was to optimize the CP layer thickness by controlling the hydrodynamic conditions (e.g., pressure and cross-flow velocity). The experiments were conducted with a bench-scale system similar to that used in the membrane performance tests. To better identify the potential foulants in the feedwater and to gain insight into the mechanisms of flux decline, a particle size analysis was carried out on the raw and filtered (with a 5-μm cartridge filter) seawater. The raw seawater consisted of particles primarily in the 1-μm size range with a small fraction in the 100-nm range. The filtered seawater only contained the particles in the 100-nm range. From the particle analysis, it appears that the 5-μm cartridge filter would be effective in removing all of the 1 μm particles.

Table ES.1
Mass transfer coefficients from synthetic seawater at 20°C (simulated Stage 1)

	k_w (gfd/psi)	P (psi)	k_{Na+} (gfd)	k_{K+} (gfd)	k_{Ca2+} (gfd)	k_{Mg2+} (gfd)	k_{Cl-} (gfd)	k_{Br-} (gfd)	k_{HCO3-} (gfd)	$k_{SO4\ 2-}$ (gfd)	k_{TDS} (gfd)
NF-90	0.088	500	1.93	2.38	0.07	0.04	1.56	2.33	0.26	0.02	1.34
		575	2.00	2.27	0.07	0.04	1.52	2.19	0.27	0.03	1.32
NE-90	0.077	500	3.69	5.34	1.72	1.32	3.68	4.26	2.02	1.11	3.34
		575	3.80	5.55	1.82	1.41	3.88	5.39	2.11	1.02	3.56

Table ES.2
Mass transfer coefficients from simulated Stage 2 experiments using Stage 1 permeate obtained with two different pressures (T = 20 °C)

Membrane	P1 (psi)	k_w (gfd/psi)	P2 (psi)	k_{Na+} (gfd)	k_{K+} (gfd)	k_{Ca2+} (gfd)	k_{Mg2+} (gfd)	k_{Cl-} (gfd)	k_{Br-} (gfd)	k_{HCO3-} (gfd)	$k_{SO4\ 2-}$ (gfd)
NF-90	500	0.239	200	2.46	2.08	ND	ND	2.08	3.84	ND	ND
			300	2.62	2.05	ND	ND	2.05	4.30	ND	ND
	575	0.242	200	1.45	1.19	ND	ND	1.35	2.39	ND	ND
			300	1.51	1.20	ND	ND	1.43	2.60	ND	ND
NE-90	500i	0.166	200	2.11	2.42	1.25	1.06	2.30	2.00	1.41	0.38
			300	2.71	3.09	1.73	1.35	3.13	2.10	1.86	0.39
	575	0.164	200	1.93	1.92	1.20	0.86	2.29	1.76	1.29	0.32
			300	2.49	2.91	1.67	1.23	2.98	2.18	1.75	0.44

Note: ND indicates that the levels of ions in Stage 2 permeate were not detectable by the methods used. Because the feed concentrations were also very low, resulting k_i's were out of range and thus were not reported
P1: pressure with which simulated Stage 1 permeate was obtained (permeate is used for simulated Stage 2 tests)
P2: pressure for the simulated Stage 2 experiments

The general fouling characteristics of the membranes were determined by evaluating flux decline over a 72-hr period for a constant applied pressure and flow rate. The flow rate of 1 L/min corresponded to the cross-flow velocity used in the membrane performance experiments. Both membranes experienced a 17-percent flux reduction over the 72-hr period, and both lost flux at a similar rate. The inability to identify a critical flux in the pressure/flux range of interest was not unexpected because of the relatively low flux decline observed in the first few hours of the experiment (as compared to other critical flux studies where substantial fouling is seen in the first hour of testing). Because of the dilute particulate concentration of the feed water, it is likely that the observed flux decline was largely due to concentration polarization and increased osmotic pressure, and thus would not readily result in a critical flux value.

The purpose of the next set of experiments was to analyze the CP layer that develops during seawater desalination by NF. Two factors affecting CP are applied pressure and cross-flow velocity (Mulder 1996). Flux and rejection decreased with decreasing cross-flow velocity because of the increased thickness of the CP layer. A higher ion concentration at the membrane surface increases osmotic pressure and particle diffusion through the membrane, thus resulting in decreased TDS rejection and permeate flux. Cross flow may not be sufficient to compensate for the increased particle loading to the surface that comes from a large initial flux.

OPERATIONAL OPTIMIZATION

A detailed plan to test various NF membranes in an 8-gpm pilot unit was developed to consider the impact of temperature, operating pressure in the first and second stages, and array configuration for concentrate recycle. Although seawater salinity remains fairly constant at about 34,500 mg/L, the temperature from the ocean near Long Beach varies between 15 and 20°C throughout the year, and was expected to affect the process performance. The pilot plant configuration is shown in Figure ES.1 below. Flow streams recycled to the feed include Stage 2 concentrate (at all times) and various amounts of the more saline part of Stage 1 permeate (e.g., 0, 20, and 40 percent of the total Stage 1 permeate flow). These recycle streams do not change the fundamental nature of the process and were included in the test plan to provide refinements that would possibly further enhance overall recovery.

Pilot Test Results

Tables ES.3 and ES.4 show representative results from the 100 pilot tests that were conducted for the FilmTec NF-90 and Saehan NE-90 membranes, which was used in calibrating the predictive model. Determining the linear dependence of the reflection and permeability coefficients on operating parameters allowed the model to predict within reasonable accuracy (±10 percent) the relationship between finished water quality, energy required (operating pressure), and recovery.

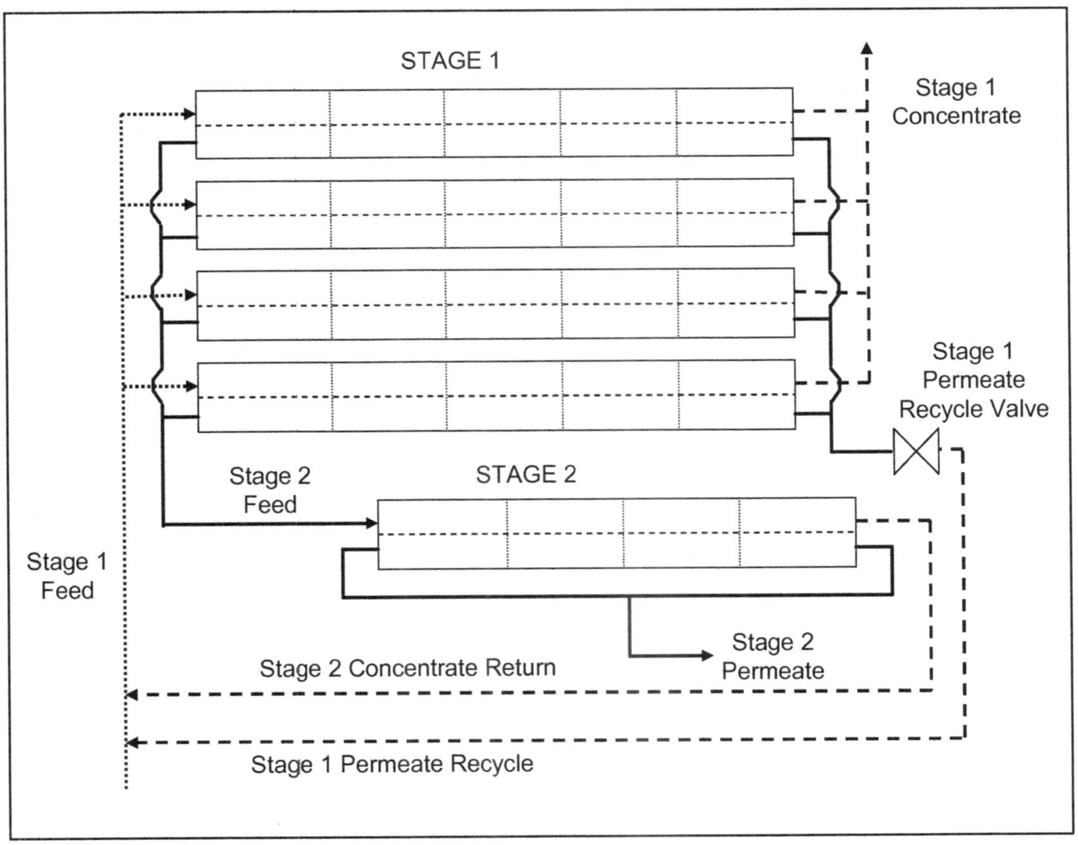

Figure ES.1 Configuration of pilot plant

Table ES.3
Selected pilot test results for FilmTec NF-90 at average temperature 67°F (19.4°C)

Stage 1 pressure (psi)	Stage 1 permeate recycled (percent)	Stage 2 pressure (psi)	Stage 1 feed salinity (mg/L)	Stage 2 permeate salinity (mg/L TDS)	Recovery (percent)	Energy (kWh/kgal)
580	0	170	28,071	180	39	16.9
		210	31,440	328	41	14.3
	40	170	29,458	176	33	16.7
		210	29,920	331	34	16.4
520	0	170	31,770	206	29	15.9
		220	34,478	380	31	13.8
	40	145	29,392	194	26	18.4
		190	30,185	349	26	16.6

Table ES.4
Selected pilot test results for Saehan NE-90 at average temperature 70°F (21.1°C)

Stage 1 pressure (psi)	Stage 1 permeate recycled (percent)	Stage 2 pressure (psi)	Stage 1 feed salinity (mg/L)	Stage 2 permeate salinity (mg/L TDS)	Recovery (percent)	Energy (kWh/kgal)
520	0	200	24,702	338	42	19.7
	40	140	25,825	176	39	19.1
		200	28,996	355	40	16.0
470	0	170	25,429	364	33	20.3
	40	125	26,631	184	33	19.8
		160	30,237	363	33	16.7

The salinity of Stage 1 feed water in the pilot tests varied from approximately 25,000 to 33,000 mg/L TDS, depending on the quantity and quality of the recycled flows (i.e., Stage 2 concentrate with or without the more saline part of Stage 1 permeate). The system provided 99 percent overall rejection of the feed salinity with the NF^2 process, and a final permeate salinity meeting drinking water standards (i.e., less than 500 mg/L TDS) could be produced without recycling Stage 2 concentrate. As demonstrated by the data in Tables ES.3 and ES.4, Stage 1 and Stage 2 operating pressures were the main factors in determining final permeate flow and salinity.

Energy requirements were similar for both membranes. Assuming a 30 percent recovery in each stage of a brine-staging SWRO system (i.e., the first-stage concentrate is treated in a second stage, and both permeates are mixed resulting in an overall 51-percent recovery), operating pressures of approximately 600 and 850 psi for the respective stages, and using other assumptions similar to those for the NF membranes (e.g., pump efficiency, energy recovery), the calculated energy required is approximately 18 kWh/kgal. This means that some conditions tested in Tables ES.3 and ES.4 did not result in energy savings as compared to single-pass SWRO.

The energy required increased as final permeate salinity decreased. In Stage 2, a higher applied pressure increased the recovery, which decreases the concentrate flow. This resulted in less feedwater to treated in Stage 1, and because Stage 1 operates under a substantially higher pressure than Stage 2 (approximately double that of Stage 2) a decrease in Stage 1 feed flow will decrease the required energy (energy calculations were based on flow and pressure).

Both NF membranes rejected a greater proportion of divalent ions compared to monovalent ions, which shifted the proportion of sodium and chloride ions from about 87 percent (weight basis) in the raw water to more than 95 percent in the final permeate salinity. The preferential removal of sulfate, magnesium, and calcium ions compared to monovalent ions in the first stage of the NF^2 process allows a relatively scale-free environment in the second stage. The concentration polarization modulus (CPM) in either stage is less than for single-pass SWRO (i.e., membrane selectivity and applied pressure are substantially greater in SWRO) and alters the supersaturation level of sparingly soluble salts so that antiscalant dosing in the NF^2 process can be substantially trimmed down from dosages typically used in SWRO.

The combination of decreased concentration polarization and preferential divalent ion removal makes the process particularly well suited for using pH elevation in the second stage to remove boron, because the scaling tendency under these conditions is lowered when compared to SWRO. Figure ES.2 shows boron removal increased with increase in pH, especially beyond pH 9.8 where boron fully hydrolyzes to borate. Approximately 80 percent of the raw water boron concentration (4.5 mg/L on average) was rejected beyond pH 10, resulting in permeate boron concentration below California's Notification Level (1 mg/L).

Mixed Membrane Tests

Pilot tests were conducted with various "looser" membranes in Stage 1 and "tighter" membranes in Stage 2 to determine if any such combination would further improve energy and/or recovery of the system. The terms "looser" and "tighter" are relatively terms refering to the molecular weight cutoff. The salinity target was 200 mg/L TDS for all tests. Table ES.5 clearly shows that increasing membrane tightness in Stage 1 (from TS-80 to NF-90) decreased the theoretical energy required. A "loose" membrane in Stage 1 lowers the permeate quality, which requires a "tight" membrane in Stage 2 to ensure good rejection. Although this system would maximize overall recovery, it would not be the most energy-efficient configuration. Conversely, using a "tight" membrane in Stage 1 may not produce enough permeate to feed Stage 2 to achieve an acceptable overall recovery and increasing the energy required.

The relationship between recovery and rejection was modeled to determine the benefit, if any, of mixing membranes with various salt rejection and flux capabilities on the overall system performance. Equations were developed so that all recoveries (Y) can be determined based on:

- Rejection characteristics of membranes in Stage 1 (R_1) and Stage 2 (R_2)
- Raw water flow and salinity
- % (percent) of Stage 1 permeate flow being returned
- Salinity for Stage 1 feed and concentrate flows

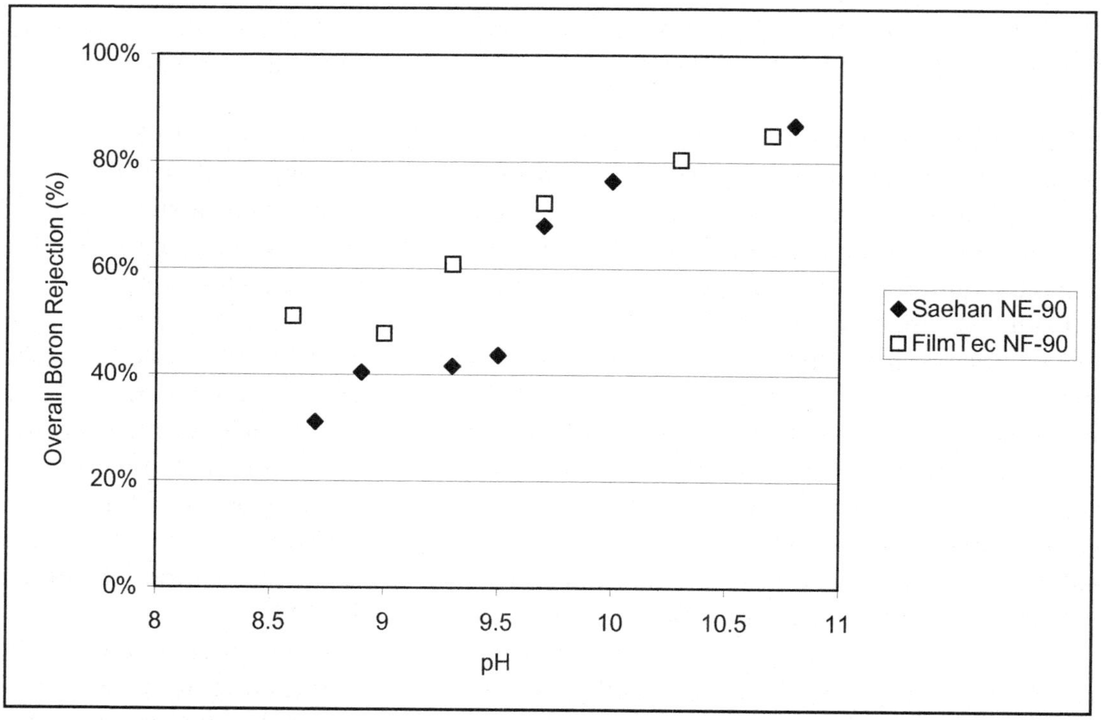

Figure ES.2 Selected pilot test results for boron removal at average temperature 70°F (21.1°C) (boron feed concentration range: 4.6 to 4.8 mg/L)

Table ES.5
Selected pilot test results for combination of NF and low pressure RO membranes

Membrane					
Stage 1	Stage 2	Stage 1 recycle (percent)	Stage 2 pressure (psi)	Overall recovery (percent)	Energy (kWh/kgal)
TS-80	ESPA4	40	190	24	40
TS-80	ACM	40	270	27	34.4
NE-90	ESPA4	40	140	41	20.5
NE-90	ACM	25	300	39	18.0
NF-90	ESPA4	0	160	28	16.9
NF-90	ACM	0	290	27	15.8

Stage 1 pressure: 580 psi, target salinity: 200 mg/L TDS

$$Y_1 = \frac{C_{b,1} - C_{f,1}}{C_{b,1} - C_{f,1}(1 - R_1)} \qquad \text{(ES.5)}$$

$$Y_2 = \frac{\left(\dfrac{R_1\, C_{f,1}}{C_{b,1} - C_{f,1}}\right)\left(C_{b,1} - C_{raw}\right) - \%\left(1 - R_1\right)\left(C_{b,1} - C_{f,1}\right)}{\left(1 - \%\right)\left(C_{raw} - C_{f,1}\left(1 - R_1\right)\left(1 - R_2\right)\right)} \qquad \text{(ES.6)}$$

$$Y_{overall} = \frac{1}{1 + \dfrac{R_1\, C_{f,1}}{\dfrac{R_1\, C_{f,1}\left(C_{b,1} - C_{raw}\right) - \%\left(1 - R_1\right)\left(C_{b,1} - C_{f,1}\right)^2}{\left(1 - \%\right)\left(C_{raw} - C_{f,1}\left(1 - R_1\right)\left(1 - R_2\right)\right)}}} \qquad \text{(ES.7)}$$

Based on results from the "optimization" model, conditions that result in low Stage 1 feed flow rate will also result in low energy required. Low Stage 1 feed flow occurs when:

- Stage 1 concentrate salinity is maximized (i.e., recovery in Stage 1 is maximized)
- Dilution of Stage 1 feed salinity from recycle streams is minimal
- Partial recycle of Stage 1 permeate is avoided
- Tighter membranes are placed in Stage 1 (94 percent rejection) and looser membranes in Stage 2 (84 percent rejection)

The first three conditions have been simultaneously pilot tested and the related low energy values shown in Table ES.3 (rows 2 and 6: 14.3 kWh/kgal at 580 psi and 13.8 kWh/kgal at 520 psi) confirms their validity.

BLENDING STRATEGIES

LBWD's existing finished water blends varies seasonally, and can range up to 100 percent of treated groundwater (GW) in summer, and up to 100 percent of purchased surface water (SW) in winter. The surface water can be a blend of State Project water (SPW) or Colorado River water (CRW). The goal of this phase was to determine the percentage of desalinated water (DW) that could be blended taking into account the issues of disinfection by-product (DBP) formation, disinfectant residual, and corrosivity. Table ES.6 presents water quality characteristics for the various waters to be blended. The characteristics for DW vary depending on the permeate salinity and pH targets. The two pHs considered represent one that is similar to the pH condition of the water in the existing distribution system (8.0), and one that is required if additional boron rejection is necessary (with an interstage pH adjustment to pH 9.8).

DBP evaluation was performed on a mass balance basis only. It is unlikely that total trihalomethanes (TTHM) or five haloacetic acids (HAA5) would be formed upon chlorination of the permeate (for primary disinfection) because most of the available total organic carbon (TOC) would be removed through the NF^2 system. However, bromide in the permeate may create a stability problem for the residual disinfectant. Bromide reacts very quickly with free chlorine to form hypobromous acid (HOBr), and represents a source of chlorine demand (AWWA 1999).

Table ES.6
Water quality characteristics of waters used for the blending evaluation

Water	TDS (mg/L)	pH	Alk (mg/L)	Ca (mg/L)	SO_4 (mg/L)	Cl (mg/L)	Br (mg/L)	LSI	RSI	CCPP
GW	240	7.8	130	25	25	35	0.05	-0.4	8.6	-4.6
SPW	330	8.2	100	35	90	55	0.16	0.0	8.3	-0.4
CRW	650	8.0	125	75	265	90	0.08	0.2	7.7	2.7
DW	200	8.0	3.8	0.9	1.1	120	0.5	-3.1	14.2	-11.0
		9.8	3.8	0.9	1.1	120	0.5	-1.3	12.4	-10.9
	350	8.0	3.7	1.5	1.8	226	1.1	-3.1	14.3	-12.0
		9.8	3.7	1.5	1.8	226	1.1	-1.3	12.5	-11.9

LSI: Langelier Saturation Index; RSI: Ryznar Saturation Index; CCPP: calcium carbonate precipitation potential

Bromide

Like chlorine, bromine (in its hypobromous form) reacts with ammonia to form bromamines (White 1986). Chloramines and bromamines will then be formed when ammonia is added to form the residual disinfectant. Dibromamine ($NHBr_2$) is very active compared to mono- and dichloramine and usually decomposes rapidly, reverting to bromide ion, which terminates the bromine residual (White 1986). Because the desalinated water will be blended with surface waters containing TOC, dibromamine (and any HOBr resulting from breakpoint bromination) will react with the TOC to form brominated disinfection byproducts.

Because chloramine does not oxidize bromide ions to bromine, the depletion of residual disinfectant will stop when $NHBr_2$ and HOBr have decomposed to bromide ions. However, monochloramine has been shown to decompose in presence of bromide ions, and disinfectant residual depletion may continue beyond the decomposition of $NHBr_2$ and HOBr, albeit at a lower rate. A possible residual disinfectant control option would be to "quench" the oxidant demand from the unreacted TOC of the surface waters with additional chlorine (which would create additional TTHM and HAA5) to prevent brominated residuals from forming brominated DBPs, and avoiding disinfectant residual depletion. The formation of additional DBPs would be balanced by the dilution factor associated with the blending.

Corrosivity

Several factors influence the rate of corrosion, including pH, temperature, and presence of dissolved solids and gases. Decreasing the system pH increases corrosion because more hydrogen ions become available. When the water temperature is increased, the rate of corrosion is generally increased. Therefore, corrosion rates in the summer are typically higher than in the winter. However, colder temperatures allow for more dissolved oxygen, which can also increase corrosion.

Many indices are available to predict scaling and corrosion tendencies of a system. The ones referenced here include the Langelier Saturation Index (LSI), Ryznar Saturation Index (RSI), and Calcium Carbonate Precipitation Potential (CCPP). The blending model was developed by enhancing a corrosion control software programs (Rothberg, Tamburini & Winsor Blending Application Package 4.0, Item number 53042 in AWWA Bookstore) with Microsoft Excel® macros to allow evaluation of multiple blends. The model can be used to determine the optimal blend of waters for minimizing the corrosion potential of the final mix, or to evaluate the possible corrosivity effects by increasing the percentage of desalinated water in the final mix. Minimizing the corrosivity of the final blend may not minimize DBP formation, and any model result based on minimizing corrosivity requires an assessment of the DBP levels in the final blend.

The resulting blend becomes more corrosive as the proportion of GW increases in relation to SW. Only CCPP and RSI appear to capture variations in SW makeup, with 100 percent SPW being more corrosive than 100 percent CRW (Figures ES.3 and ES.4). The CCPP is affected by the permeate salinity, with 350 mg/L TDS being more corrosive than 200 mg/L TDS for the same pH values (Figure ES.5). Of the indices considered, it appears that CCPP shows the greatest sensitivity to the various water quality variations, and is therefore preferable to the other indices. For the summer condition, blending DW at pH 8.0 with GW would likely lead to corrosive conditions. In winter, corrosive conditions would exist if the proportion of DW at pH 8.0 is too high as compared to SPW.

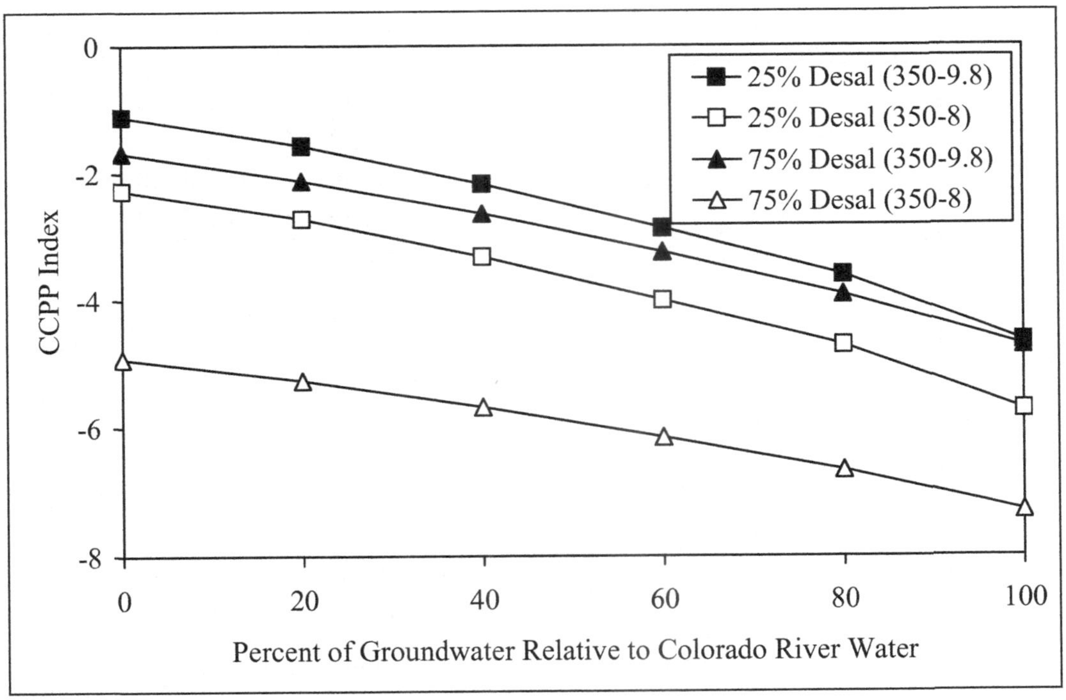

Figure ES.3 Effect of surface water composition on CCPP for desalinated water (at 350 mg/L TDS and pH 8 or 9.8) blended with groundwater and Colorado River water

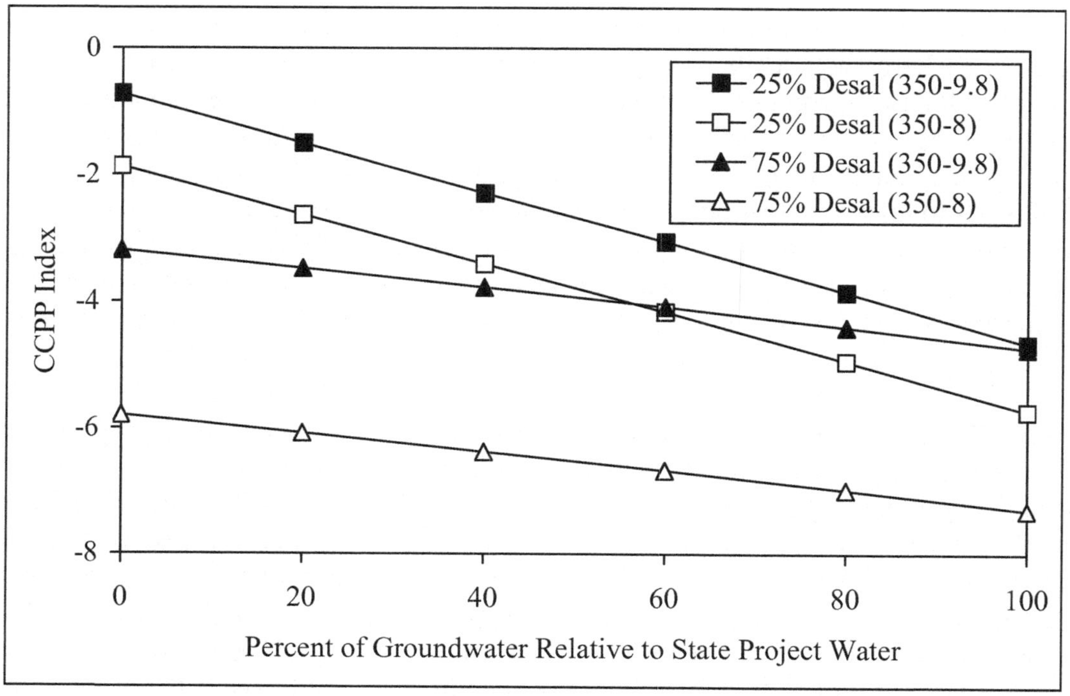

Figure ES.4 Effect of surface water composition on CCPP for desalinated water (at 350 mg/L TDS and pH 8 or 9.8) blended with groundwater and State Project water

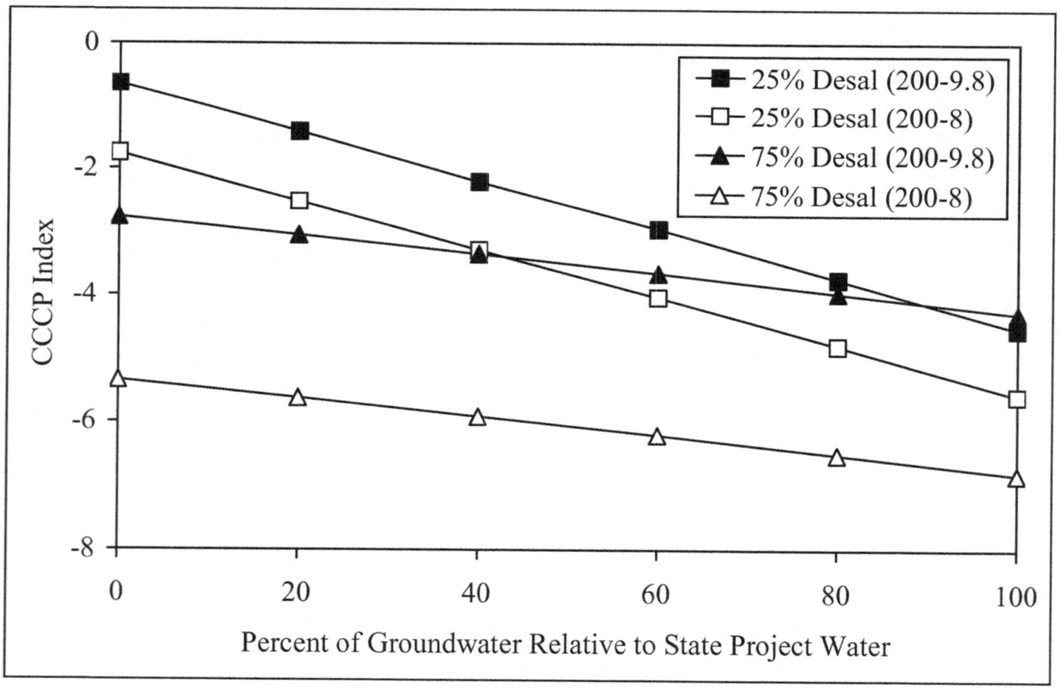

Figure ES.5 Effect of surface water composition on CCPP for desalinated water (at 200 mg/L TDS and pH 8 or 9.8) blended with groundwater and State Project water

Based on these findings, LBWD has the following available options to mitigate corrosion potential from blending DW with existing finished water:

- Increase corrosion inhibitor dosage (e.g., orthophosphate) in summer when GW is used, as well as in winter when SPW is used for SW
- Maintain the elevated pH necessary for the boron removal step in DW (the impact of high pH on the chloramination of DW will need to be assessed)
- Maintain DW in the final blend at 25 percent or less in winter if DW needs to be at pH 8.0

VIRAL CHALLENGES

Although NF is a non-porous semi-permeable membrane that represents a barrier to viruses, the associated modules are generally not manufactured to be aseptic. It is clear that a disinfection step should be incorporated in the future seawater desalination system, even if pathogen removal by the NF^2 process is greater than the minimal regulatory goal as required by the SWTR. The original scope of work included a viral challenge phase, which was intended to provide preliminary data on the inherent redundancy of the system and the possible impact that recycling streams would have on virus accumulation. The main findings would form the basis for a more complete virus challenge study that may be required by the California Department of Health Services (CDHS).

MS-2 bacteriophage was the virus selected for the tests because it has become an accepted surrogate in the water industry. However, preliminary recovery tests of MS-2 inoculated in a seawater matrix resulted in non-detect for the male specific (MS) titer and inconsistent recoveries for the somatic titer. Because of the analytical issues associated with these tests, even under laboratory conditions, the research team determined that running pilot viral challenge tests using MS-2 coliphages in seawater would not be meaningful, especially when log-removal calculations involve phage quantification.

Non-viral alternative surrogates were considered in lieu of MS-2 bacteriophages for the tests, but each had drawbacks (fluorescent microspheres would have exceeded the budget; molecular markers, starch, and tracer dyes would have irreversibly fouled the membrane). Because the objective of the project was to apply an existing analytical method to determine the system removal and not to develop a new analytical method for viruses in a seawater matrix, the challenge test was eliminated. However, it is important to consider viral surrogates in addition to MS-2 phages for regulatory validation of a seawater desalination process.

CONCLUSIONS AND RECOMMENDATIONS

The Long Beach method of using dual-staged nanofiltration (NF^2) can desalt seawater to potable water levels (i.e., less than 500 mg/L TDS) with 20 percent less energy than is theoretically needed for single-pass SWRO (at 18 kWh/kgal), and at a recovery of approximately 40 percent (typical SWRO system recovery is 50 percent). Additionally, permeate boron concentration achieved from this novel process when Stage 2 is at pH 10 is below California's boron Notification Level (1 mg/L).

Suitable Membrane

Pressure applied in Stage 2 directly affects the final salinity, and because the primary salt in seawater is NaCl, it is essential that the NF membrane selected for seawater desalination exhibit high rejection for this salt. Optimization model results show that a membrane rejecting 90 percent sodium chloride (NaCl) at ionic strength similar to seawater at 20°C would be suitable if the permeate flux is satisfactory (e.g., greater than 15 gfd).

It was shown that a low salinity target (e.g., 200 mg/L) requires more energy than a higher salinity goal (e.g., 350 mg/L). When compared to looser NF membranes, tighter NF membranes were demonstrated to require less incremental change in overall energy input as the final salinity target changed from 350 to 200 mg/L TDS. To compensate for the lower salt rejection, the looser membranes depend more on the recycle stream in order to dilute the feed water to achieve the final permeate salinity target, and therefore more flow needs to be pumped, which increases energy consumption. Membrane selectivity becomes essential in optimizing energy required; it was observed that a tighter membrane provides greater membrane resistance, which requires additional pressure and energy, and a loose membrane requires more flow to be processed in order to achieve the same water quality target through dilution of the feedwater. For tighter membranes, recycling flows to Stage 1 feed is of little value and may actually increase the energy required per volume of water treated.

Process Operation

Pressure in Stage 2 should be set to maximize recovery in Stage 2 while not impacting final permeate quality. Overall system recovery is improved at applied pressure in Stage 1 close to manufacturer specified limits (typically 600 psi). Because Stage 1 contributes 80 percent of the total energy required, any efforts at energy savings should focus there. The preference for divalent ion removal in the first stage alters the supersaturation conditions of sparingly soluble salts so that the process is well-suited for boron control by increasing pH in the second stage.

System Integration

The two-pass NF system removed most of the seawater TOC, and the formation of TTHM and HAA5 after primary disinfection of the permeate with chlorine should be minimal. However, the permeate will include bromide ions, which will react with chlorine to form bromine and hypobromous acid (HOBr). Two issues are thus associated with bromide ion presence in the permeate:

- Bromide exerts chlorine demand, therefore the chlorine dose needs to be increased to meet C·T requirements
- HOBr reacts with ammonia to form bromamines, and unlike chloramines, will react with TOC when desalinated water is blended with surface water. Brominated DBPs will be formed in this process, before reverting to bromide ions. Disinfectant residual depletion will slow down after all bromamines and HOBr have decomposed to bromide ions (chloramines do not oxidize bromide ions to bromine).

NF^2 Prototype Testing

Operating conditions provided by the predictive model will guide the long-term testing of the NF^2 process with a 300,000-gpd prototype, which will include a conventional SWRO train for side-by-side comparison. Also, the predictive model suggested that most of the desalination takes place with the first four elements in Stage 1 and the first three elements in Stage 2 will have to be verified to further optimize the process.

The relationship between recovery (water quantity) and rejection (water quality) was modeled to determine the benefit of mixing membranes with various salt rejection and flux characteristics on the overall system performance. Conditions that result in low energy input included using tighter membrane in Stage 1 (94 percent rejection) and looser membrane in Stage 2 (84 percent rejection). This suggests that the membrane configuration may be an area of research for the prototype plant

It was observed that rejection decreased in bench and pilot tests after a critical pressure was exceeded for some salts. Convection dominated and permeate water quality improved with increasing pressure before this critical point. Diffusion dominated and permeate water quality deteriorated with increases in pressure after the critical point was reached. This phenomenon will be verified at the demonstration-scale to determine the balance of water quality and energy input.

As shown in this report, an increase in cross-flow velocity improved flux and rejection because the CP layer thickness was decreased. However, increasing cross-flow velocity may not be sufficient to overcome the increased particle loading to the membrane surface, which results from a high initial flux. Therefore, various loading rates on the membranes should be evaluated during prototype testing for their ability to mitigate membrane surface fouling. The loading rates can be varied by increasing raw water flow or modifying the number of parallel Stage 1 pressure vessels.

Disinfectant Residual Stability

Additional investigation is needed to address the residual disinfection issue. In particular, the impact of maintaining a high pH for corrosion control purposes (mentioned above as an option) on chloramination needs to be assessed. Controlling the effects of bromination (i.e., brominated residuals forming brominated DBPs and depleting disinfectant residual) will be essential for system implementation. A study could investigate mitigation strategies including "quenching" the oxidation demand from available TOC of the surface waters with a chlorine addition (the additional TTHM and HAA5 formation could be minimized by dilution through blending).

Corrosion Study

Because only a desktop corrosion study was performed in this study, blending strategies and the related corrosion issues will need to be investigated further in a year-long pipe-loop study (to cover the seasonal variations in water blends), which will should include actual pipe samples from Long Beach's distribution system. Various permeate conditions (e.g., salinity, pH) resulting from prototype plant tests will be used to make up various blends.

Viral Surrogates

The poor recovery of MS-2 phages in seawater matrix indicates that other surrogates are needed. Suitable virus surrogates would be less than 30 nm and/or molecular weight less than 100,000 amu. Additionally, a conservative challenge particulate should have a neutral surface charge, because charged particles may interact with other particles and surfaces, thus enhancing removal. The surrogate should also be cost effective, easy to use, and measurable.

CHAPTER 1
THEORETICAL BASIS

BACKGROUND

Reverse osmosis (RO) is a proven membrane technology for seawater desalination. Although costs have been significantly reduced, the energy required for seawater desalination is high as compared to the energy required for treatment of other source waters (e.g. surface or ground water) due to the high pressures required (typically from about 800 to 1,000 psi). An alternative to traditional seawater RO (SWRO) is a novel dual-staged nanofiltration (NF^2) process developed by the Long Beach Water Department (LBWD). Although NF membranes have been primarily used for hardness removal, and disinfection by-product (DBP) precursor removal, this novel NF system is able to desalt seawater by treating the first stage permeate through a second stage in order to produce finished water with salinity levels that meet drinking water standards. Beyond actual seawater desalting, the NF^2 process has additional benefits compared to single-pass SWRO, namely:

- Two physical barriers:
  - Provide system redundancy for accrued water quality protection
  - Allow boron removal through pH increase in the second stage
- Lower pressure requirements leading to energy savings

The comparative looseness of NF membranes means that the membrane hydraulic resistance should be less than for SWRO. This increased water permeability (water produced per membrane area [flux] for a given applied pressure) in NF membranes optimizes the operating pressure by decreasing the energy penalty associated with membrane resistance (i.e., the same amount of permeate flow can be obtained with NF compared to SWRO at an applied pressure closer to the feed osmotic pressure).

Figure 1.1 plots minimum operating pressures required to overcome the osmotic pressure of seawater at specific recoveries (recovery is the ratio of water produced by the system to water fed into the system). The graph also shows the behavior of the energy required to produce 1,000 gallons of permeate at the same recoveries. Figure 1.1 is based on the following assumptions:

- Salinity is related to recovery using mass balance (see Eq. (1.10))
- Initial seawater salinity of 35,000 mg/L TDS
- Membrane system providing 99 percent salt rejection
- Osmotic pressure empirically related to salinity as (psi) = 0.011 × (mg/L TDS)
- Energy per volume of permeate is calculated as the product of pressure and the inverse of recovery and using an energy efficiency of 75 percent

Figure 1.1 shows that even though the required pressure increases with recovery, the energy per water produced follows a concave curve with a minimum at 50 percent recovery. Therefore energy considerations need to go beyond system pressure and include recovery. In addition, the graph neatly illustrates the point that for seawater desalination efficiency (i.e., high recovery) does not entail affordability.

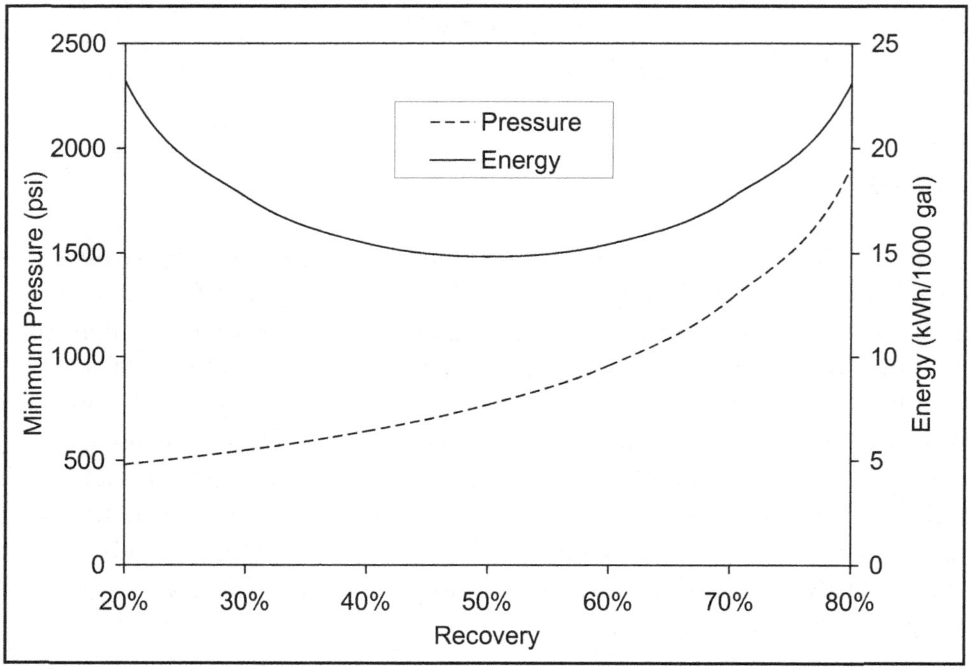

Figure 1.1 Minimum pressure required and energy per water produced at various recoveries for seawater desalination by membrane processes

The NF^2 process is more than a double-pass NF system because it recycles the concentrate flow from the second pass (and, if desired, the concentrate and part of the permeate from the first pass) back to the feed. Hence, LBWD's NF^2 process effectively recirculates the feed several times through each NF pass, and therefore is more accurately described as dual-staged NF (permeate staging). The capability of NF^2 to reduce operational costs for seawater desalination was assessed through a research plan consisting of four phases (Theoretical Basis – Chapter 1, Operational Optimization – Chapter 2, Blending Strategies – Chapter 3, and Viral Challenges – Chapter 4). A computer model was developed as well to guide operational tests for a 300,000-gpd demonstration plant on-line in 2005.

LITERATURE REVIEW

Boron

Boron is found widely in the environment both from naturally occurring and anthropogenic sources (Simonnot et al. 2000). In seawater, boron levels of approximately 4 to 5 mg/L occur mainly from the washing of geological materials (Pastor et al. 2001). Levels of boron in desalinated seawater have become an issue because of possible developmental and teratogenic effects (Magara et al. 1998). Due to these negative health effects, the World Health Organization (WHO) has set a boron standard at 0.3 mg/L (Melnik et al. 1999). This standard has resulted in recent research on boron removal from desalinated water.

Several options are available for lowering levels of boron in drinking water including blending schemes (Simonnot et al. 2000), ion exchange (IX) (Nadav 1999, Badruk et al. 1999), adsorption (Choi and Chen 1979), and membrane filtration processes (Magara et al. 1998, Pastor et al. 2001, Prats et al. 2000). At the pH of natural water (pH 6 to 8), boron removal efficiency by RO is low. At this pH, boron is found as boric acid (H_3BO_3), which diffuses easily through RO membranes (Pastor et al. 2001). However, boron exists in the form of borate ion ($H_2BO_3^-$) at pH 9.5 and above, which is highly rejected by RO membranes.

This observation implies that operation of RO or NF membranes at high pH values is necessary for boron removal. However, raising the pH of the seawater feed would result in scaling. Thus, in order to remove boron, an additional process may be required. This can be either another RO stage at a higher pH or another process such as IX. A three-stage RO system for seawater desalination and boron removal has been investigated in several studies (Magara et al. 1998, Pastor et al. 2001, and Prats et al. 2000). The first two stages (with no pH adjustment) are dedicated to seawater desalination and the third stage (with elevated pH) is for boron removal. From these studies, raising the pH to a range of 9.5 to 10.5 in the third stage results in nearly 100 percent boron rejection (Magara et al. 1998, Prats et al. 2000, and Pastor et al. 2001).

Recently, studies have been performed to test the effectiveness of RO membranes developed specifically for high boron removal. For example, Taniguchi et al. (2004) studied the TM820A SWRO membrane (Toray, Shiga, Japan). This membrane has a reduced affinity for boron and a tighter molecular structure in its active layer to provide better size exclusion of boric acid. Results showed that the membrane provided 94 to 96 percent boron rejection. Instead of adding another RO stage at higher pH, IX has been evaluated for treating seawater RO permeate. Nadav (1999) tested Amberlite IRA 741 (Rohm and Haas, Philadelphia, PA) IX resin in pilot scale studies. This resin effectively removed boron at a reasonable production cost (Nadav 1999). No literature has been found describing boron removal by NF membranes.

Disinfection Byproducts

Disinfection is required for desalinated seawater to prevent contaminants from entering the distribution system due to membrane tears or leaks and to ensure the presence of disinfection residual in the distribution system (USBR 2003). However, disinfectants can react with natural organic matter (NOM) present in the water to produce disinfection byproducts (DBPs). Possible DBPs that can be formed include trihalomethanes (THMs), haloacetic acids (HAAs), haloacetic nitriles, cyanogen halides, halopicrins, haloketones, haloaldehydes, and halophenols (Singer 1994, Kampioti et al. 2002). In all, more than 500 DBPs have been identified. DBP formation is of concern because some DBPs have been found to be carcinogenic and teratogenic (Sketchell et al. 1995, Batterman et al. 2000, Chae et al. 2002).

NOM concentrations in seawater vary depending on planetary location and ocean depth (Paytan 2004). The feedwater to the Long Beach desalination pilot plant has a TOC concentration of 0.5 mg/L (Andrews-Tate, personal communication, July 19, 2004). NF has been shown to be highly effective at removing the organic fraction of DBP precursors (Allegeier and Summers 1995, Blau et al. 1992, and Chellam 2000). NOM is removed by NF membranes through a combination of size exclusion and electrostatic repulsion (Lee et al. 2004).

Brominated Disinfection Byproducts

Bromide ion (Br^-) is of special concern because many brominated and polysubstituted (chloro- and bromo-substituted species) DBPs are suspected carcinogens. The typical concentration of bromine in seawater is approximately 65 to 67 mg/L (Holmes-Farley 1998) and the typical molar ratio of bromide to chloride is approximately 0.0039 (Kampioti et al. 2002).

When chlorine is used as the disinfectant, free chlorine rapidly oxidizes bromide to hypobromous acid (HOBr). NOM will react with the HOBr and residual chlorine to produce THMs, HAAs, and other halogenated DBPs. The ratio of HOBr to HOCl is an important factor in determining the type of DBPs formed. THM speciation shifts from chlorinated to mixed bromo/chloro to brominated with increasing bromide concentration (Chang et al. 2001).

Because NF membranes typically reject higher levels of NOM than bromide (Allegeier et al. 1995, Chellam 2000), NF membranes will not only decrease the concentration of DBP precursors, it will also change the DBP speciation after disinfection (Chellam 2000). An increase in the Br^-/TOC ratio will increase the levels of brominated DBPs. However, the overall levels of DBPs will be reduced.

The extent to which THMs and HAAs become brominated can be calculated using the bromine incorporation factor (Chellam 2000). This factor represents the molar concentration of bromine in the total THMs (TTHMs) and can be calculated using

$$n = \frac{CHBrCl_3 + 2CHClBr_2 + 3CHBr_3}{totalTHMs} \tag{1.1}$$

where [$CHBrCl_3$] is the concentration of dichlorobromomethane, [$CHClBr_2$] is the concentration of chlorodibromomethane, [$CHBr_3$] is the concentration of bromoform, and $0 \leq n \leq 3$. When n=0, $CHCl_3$ is the only THM formed, and when n=3, $CHBr_3$ is the only THM formed.

When ozone is used as the disinfectant, it reacts with bromide to produce either HOBr or hypobromite ion (BrO^-). The HOBr can react with NOM to produce brominated DBPs or it can be further oxidized by ozone to produce bromate (BrO_3^-), a probable human carcinogen. As with DBPs formed by chlorine, DBP formation by ozone increases with increasing NOM concentration. Also, increasing bromide concentration will increase the bromide speciation in DBPs, including bromate (Meijers et al. 1995). Increasing pH during ozonation will also increase bromate formation. Shukairy et al. (1994) found that bromate formation is an issue for waters containing high levels of bromide (like seawater), even at low ozone concentrations. This may be a concern because NF has not been shown to reject high levels of bromide (Allegeier et al. 1995).

Iodinated Disinfection Byproducts

In seawater, iodine concentrations can be up to 50 μg/L. If the iodine is not removed during treatment, iodinated disinfection byproducts (IDBPs) can form during disinfection. Incorporation of one or more iodine atoms into a THM molecule can result in the formation of iodinated THMs (I-THMs). These occur because iodide atoms quickly oxidize to HOI in the presence of ozone, chlorine, or chloramines (Bichsel et al. 2000). HOI can be further oxidized to IO_3^-. Iodate is desirable because it is not considered a carcinogenic risk like bromate, because ingested IO_3^- is quickly reduced back to iodide in the body, which is not considered problematic.

When ozone is used as a disinfectant, the oxidation of iodide occurs very quickly. The reaction produces OI^- and HOI species, although only a small fraction of OI^- is typically present. HOI and OI^- can either be further oxidized to IO_3^- or react with NOM to form IDBPs. However, the lifetime of the HOI and OI^- species in ozonation is very short (on the order of milliseconds) so that IO_3^- is the primary product.

Although chlorine is capable of oxidizing HOI to IO_3^-, the reaction does not occur as quickly as it does with ozone. Therefore, there is a longer time for the HOI to react with NOM to produce IDBPs. There can also be competition between chlorine molecules and iodine molecules for incorporation into THMs; this can shift the products from CHI_3 to mixed I-THMs.

Chloramines have less chemical reactivity than chlorine and are not capable of oxidizing HOI to IO_3^-. Therefore, HOI has a long lifetime, which leads to an increased reaction time with NOM and increased IDBPs. NOM influences the type of IDBP formed. As previously stated, NF has been proven to remove NOM effectively. However, no studies were found describing the removal of I^- by NF, or how this will affect the I^-/NOM removal ratio and IDBP formation/speciation.

Algal Toxins

Harmful algal blooms are events that are caused by microalgae and have a negative effect on human activity (Zingone and Oksfeldt 2000). The harmful microalgae species are predominantly dinoflagellates, diatoms, and cyanobacteria. These algae can harm surrounding marine life by producing toxins. During an algal bloom, toxin concentrations can rapidly increase in shellfish to levels that are harmful to people. Shellfish are especially affected because of the process in which they filter their food out of the water. The toxins, known as shellfish toxins (or phycotoxins), have low molecular weights (200 to 3500 daltons) and are non-proteinacious. Their chemical structures, solubilities, and modes of action are extremely varied. Other variabilities include their chemical stability with respect to pH, dissolved oxygen concentration, temperature, and light (Garthwaite 2000).

Amnesic shellfish poisoning (ASP) is caused by domoic acid, a water soluble toxin. Domoic acid is the main phycotoxin of concern for the west coast of the United States (i.e., California, Oregon, and Washington). It is produced by the diatom *Pseudo-nitzschia pungens*, which is typically 75 to100 μm long and at least 3 μm wide. The maximum permitted level of domoic acid in shellfish is 20 mg/L. Domoic acid accumulates in the tissue of the shellfish and can be harmful to humans through shellfish consumption. However, there is insufficient evidence as to whether domoic acid can be taken up directly from water (Mos 2001). Additionally, no studies have been found to directly measure domoic acid rejection by membrane processes.

Concentration Polarization, Fouling, and Critical Flux

Concentration Polarization

A major limitation of membrane processes is reduction in permeate flux and quality with respect to time. Typically, flux decreases rapidly in the initial stages of filtration and more gradually in the later stages (Field et al. 1995). The two main causes of this loss in membrane performance are concentration polarization (CP) and fouling, which have both been studied extensively. This review is meant to be a general overview of fouling in NF processes and how the concept of critical flux can help reduce the effects of CP and fouling.

CP occurs when the membrane rejects solutes, which causes an accumulation at the surface. The increase in concentration at the surface generates a diffusive flow back to the bulk feed. At equilibrium, the force on the solute towards the membrane (i.e., convective flow) is balanced by the forces on the solute away from the membrane (e.g., back diffusion) (Mulder 1996 and Schafer 2001) so that a greater solute concentration exists at the membrane surface than in the bulk feed solution (Figure 1.2).

CP has several consequences on NF processes. The flux will decrease because flux is a function of osmotic pressure, which is increased at the membrane surface (Mulder 1996). CP can also lead to solute adsorption and precipitation if the concentration at the membrane surface exceeds the solubility limits, and both of these will promote fouling (Sablani et al. 2001). It is also important to note that the increase in ionic strength that accompanies CP raises the precipitation threshold of sparingly soluble salts above their theoretical supersaturation. Additionally, the retention of low molecular weight solutes such as salts will be lower because of the higher salt concentration at the surface. Optimizing operating conditions such as crossflow velocity and applied pressure can help minimize CP (Manttari and Nyström 2000).

Mass transport across the CP layer has been investigated for flat sheet, tubular, and spiral wound membrane configurations. Yang et al. (2003) studied CP in spiral wound NF modules for NaCl feed solutions. It was found that the solute concentration at the membrane surface increases along with the crossflow velocity, but can be effectively reduced with feed spacers.

Membrane Fouling

Membrane fouling can be caused either by deposition of a layer onto the membrane surface or by blockage of the pores. Fouling causes decreased flux, and depending on the nature of the foulant, either increased or decreased rejection.

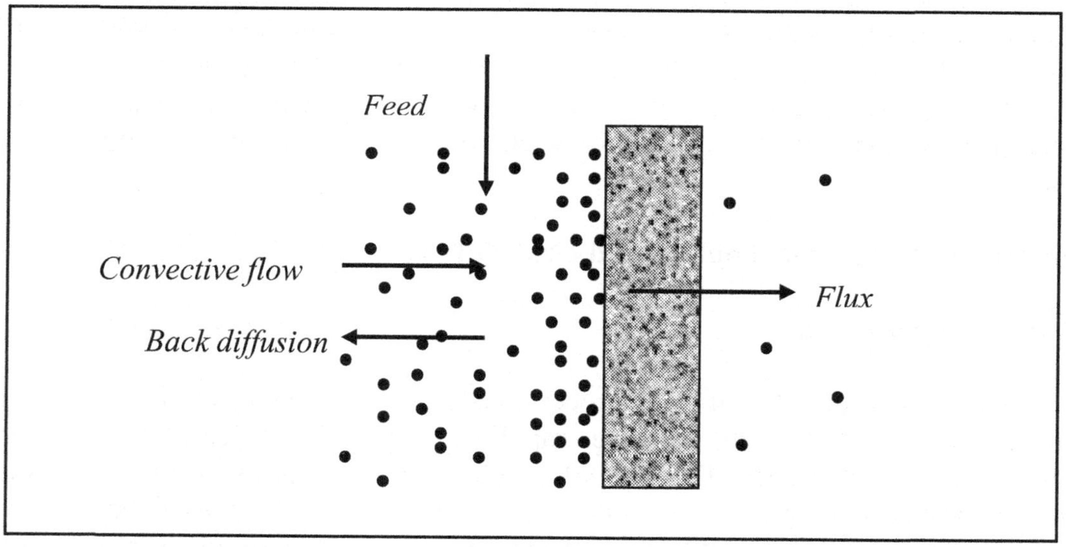

Figure 1.2 Schematic representation of concentration polarization

Flux reduction can be described by:

$$J = \frac{TMP}{\mu(R_m + R_f)} = \frac{TMP}{\mu(R_m + R_{ir} + R_r)} \tag{1.2}$$

where J is permeate flux, TMP is transmembrane pressure, μ is water viscosity, R_m is clean membrane hydraulic resistance, R_f is resistance due to concentration polarization and fouling effects, R_{ir} is irreversible fouling resistance (which is not reduced to zero when the pressure is released), and R_r is reversible fouling and polarization resistance.

Fouling in NF treatment of surface waters can occur through NOM adsorption, colloidal fouling, inorganic precipitation (i.e., scaling), and biofouling (Baker et al. 1995). NF membranes remove NOM through size exclusion and electrostatic repulsion, because NF membranes are generally negatively charged (Lee et al. 2004). NOM can accumulate at the membrane surface, creating a fouling layer that increases hydraulic resistance (Braghetta et al. 1998, Seidel and Elimelech 2002). Fouling by NOM is a function of crossflow velocity, applied pressure, and solution chemistry. As crossflow velocity increases, the shear rate also increases and reduces NOM accumulation at the membrane surface (Seidel and Elimelech 2002). Ionic composition affects the nature of NOM fouling. For example, Seidel and Elimelech (2002) found that calcium ions complex with the acidic functional groups of NOM and create a compact fouling layer. This occurs because calcium ion complexation reduces the charge of NOM and lessens the electrostatic repulsion between the membrane surface and NOM.

Hoek et al. (2002) evaluated the concept of "cake-enhanced osmotic pressure" as the cause for flux decline in the presence of colloidal foulants. In cake-enhanced osmotic pressure, the colloidal cake layer causes increased osmotic pressure at the membrane surface, which in turn causes flux decline and hinders the diffusion of salt away from the membrane surface. The large increase in salt concentration at the membrane surface can also decrease salt rejection (Lee et al. 2004).

Inorganics can foul the membrane by precipitation and scale formation if the sparingly soluble salt concentrations exceed their solubility limits. The result of scaling is flux decline (Schafer 2001). Common ions of concern for salt scaling are calcium, magnesium, and other metals (Mallevialle et al. 1996). The main cause for salts to exceed their solubility limits is through CP, and therefore minimizing CP becomes an important operational parameter.

Membrane biological fouling occurs through the formation of a biofilm. Microorganisms accumulate on the surface through attachment and growth (Mallevialle et al. 1996). Biofilms increase CP and the pressure drop through the membrane module (Mallevialle et al. 1996).

Critical Flux

Four approaches are employed to minimize fouling effects (Field et al. 1995):

1) Pretreatment to remove possible foulants prior the membrane process
2) Hydrodynamic modifications to change the flow regime
3) Membrane surface modifications to change the surface/foulant affinity
4) Cleaning to remove reversible foulants.

Recently, the concept of critical flux has been examined as an alternate method to reduce flux loss. The traditional hypothesis of critical flux is that upon startup, there exists a flux below which fouling does not occur (Field et al. 1995 and Schafer 2001).

Two types of critical flux have been identified: the strong form and the weak form. The strong form of critical flux is the flux which is equivalent to the clean water flux at the same pressure (Field et al. 1995). At the point of strong critical flux, a balance exists between forces transporting the particle or colloid to the membrane surface (e.g., convective forces) and the forces transporting the particle away from the membrane (e.g., lateral migration and shear diffusion). Therefore, no deposition of particles or colloids is occurring on the membrane surface. The definition of weak critical flux states that on start-up there is a constant flux that is rapidly established and can be maintained (Field et al. 1995). This flux does not have to be equal to that of clean water. Critical flux is a function of the ratio of particle to pore size, the system hydrodynamics, and the membrane-colloid interactions (Schafer 2001). Critical fluxes have been shown to exist for colloids, mineral suspensions, proteins or yeast, and activated sludge (Manttari and Nyström 2000).

The critical flux for NF membranes is dependent upon several parameters including cross-flow velocity, colloid feed concentration, pH, particle size, ionic strength, membrane pore size, temperature, and membrane materials. Increasing cross-flow velocity results in increased mixing and a larger diffusion away from the membrane surface, thereby resulting in a higher critical flux (Manttari and Nyström 2000). Increasing the colloid feed concentration decreases the critical flux. A higher colloid concentration causes a larger concentration polarization layer, which restricts flow and decreases the critical flux (Manttari and Nyström 2000). NF membranes are usually negatively charged, with the charge becoming increasingly negative at higher pH values. If colloids present in the feed water are negatively charged, electrostatic interactions will be repulsive. Thus, critical flux increases with increasing pH because the interactions become more repulsive (Manttari and Nyström 2000). The opposite should occur for positively charged particles because the interactions are attractive, creating a cake layer.

The original concept of critical flux described above has been expanded to become a theoretical, experimental, and operational concept. From a theoretical standpoint, the definition has remained similar to that above with a critical Peclet number (i.e., the flux divided by the mass transfer coefficient, J/k) describing the point where there is no net force on a particle towards the membrane, and a critical concentration for the flux at which the particle/surface interactions start to become irreversible (Howell 2004 and Bacchin et al. 2002).

For the experimentalist evaluating critical flux for a defined system, several transitions become important. The linearity transition is shown in Figure 1.3 and occurs when pressure is no longer linearly proportional to flux, which indicates that flux decline is occurring (Wu et al. 1999). The irreversibility transition occurs when fouling cannot be reversed by reducing flux (Espinasse et al. 2002). This transition is measured in a flux stepwise manner: pressure is increased and then decreased to the previous setting to determine if the same flux is observed. If the previous flux cannot be achieved, then the deposition is no longer reversible (Wu et al. 1999, Espinasse et al 2002, and Matsammuuronen et al. 2002). The deposition transition, which occurs at a flux where particle deposition just begins, is measured through direct observation of the membrane surface (DOTM) during fouling experiments (Schafer 2001 and Li et al. 2000).

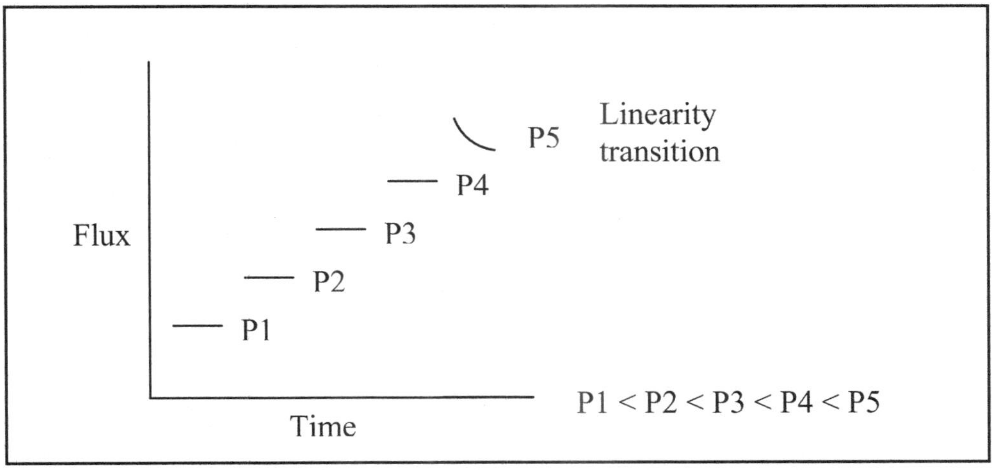

Figure 1.3 Example experimental results for determining the linearity transition

These concepts are defined for ideal systems that can be strictly controlled. However, this is not the case in real systems where a complex mix of solute membrane interactions is occurring. To apply the critical flux concept to this type of system, a sustainable flux policy has been developed (Howell 2004). Under this policy, the system is operated in a cycle of operation and cleaning so that irreversible fouling does not occur. A sustainable flux policy tries to balance the trade-offs between capital cost (e.g., membranes) and operating cost (e.g., cleaning chemicals). This is done by operating at an optimized flux that balances the flux required to meet the demand with the cleaning required to remove fouling (Howell 2004).

MEMBRANE CHARACTERIZATION

Several commercially-available NF membranes were selected for this study. These included Filmtec (Midland, MI) NF-90, TriSep (Goleta, CA) TS-80, and Saehan (Seoul, Korea) NE-90. These membranes were selected upon their designation as NF membranes by the manufacturer and their high salt rejection. Table 1.1 shows the rejection and flow characteristics for each membrane. All membranes had a maximum operating pressure of 600 psi and were supplied as flat sheets and stored in doubly deionized water (DDW) at 5°C. The zeta potential, contact angle, and roughness of each of the membrane surfaces were evaluated using an electrokinetic analyzer, a goniometer, and an atomic force microscope (AFM), respectively. The intent was to relate membrane surface properties to performance to develop a selection protocol.

Electrokinetic Measurements

Membrane zeta potential was determined using a streaming potential analyzer (ZetaCAD, CAD Instrumentation, Les Essarts Le Roi, France). Measurements were conducted using a 10 mM KCl electrolyte solution at 25°C over the pH range of 3 to 11. The electrolyte solution was prepared by adding KCl to DDW and adjusting the pH with either HCl or NaOH. Zeta potential was calculated from the measured streaming potential using the Helmholtz-Smoluchowski equation with the Fairbrother and Mastin substitution. A detailed description of the measurement procedure and zeta potential calculation can be found elsewhere (Childress and Elimelech 1996, Childress and Elimelech 2000).

Table 1.1
Rejection and flow characteristics reported by membrane manufacturers

Parameter	Solute	FilmTec NF-90	Saehan NE-90	TriSep TS-80
Feed Concentration (mg/L)	NaCl	2,000	----	----
	$MgSO_4$	2,000	2,000	2,000
Applied Pressure (psi)	NaCl	70	----	----
	$MgSO_4$	70	150	110
Solute Rejection (percent)	NaCl	85-95	----	----
	$MgSO_4$	>97	99.0	99.0
Permeate Flow (gpd)	NaCl	1,500	----	----
	$MgSO_4$	1,900	2,600	2,000

Note: Each membrane element is 4 in. in diameter, 40-in. long, and an approximate surface area of 80 ft^2

Membrane zeta potential as a function of pH is shown in Figure 1.4. All membranes showed the same general trend: as pH increased, the zeta potential became more negative. All membranes, except the NF-90, had a negative zeta potential throughout the pH range tested. There is only about 10 percent difference in the zeta potential values for the three membranes. For this reason, any differences in performance based on charge differences would be expected to be subtle, if noticeable at all. In general, membranes that are more negatively charged are anticipated to have higher levels of rejection for anions (e.g., bromide and iodide) than membranes less negatively charged because of electrostatic repulsion between the membrane surface and the ions. However, negatively-charged membranes may also attract cations that could form a bridge between the membrane surface and negatively-charged contaminants. Actual rejection was determined in performance tests described in a later section.

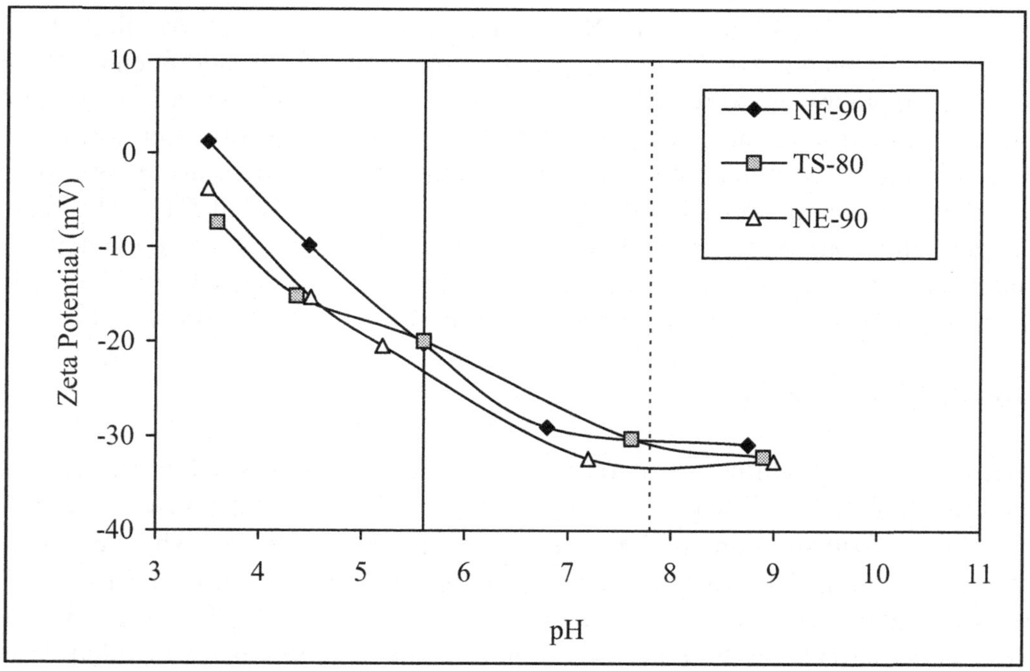

Figure 1.4 Membrane zeta potential as a function of pH (I = 0.01M KCl; T = 25°C)

Surface Energy Measurements

Contact angle measurements were performed using an NRL Contact Angle Goniometer (Rame Hart, Mountain Lakes, NJ). The goniometer can measure contact angle using both the sessile drop and captive bubble methods. In the sessile drop method, a drop of probe liquid is placed on a dry surface; in the captive bubble method, the surface is placed in a quartz cell, which contains the probe liquid, and an air bubble is released onto the surface. The captive bubble method was used to measure membrane contact angle because of its advantages over the sessile drop technique (Kwok et al. 1996).

A membrane coupon having the approximate dimensions of 1.0 in. × 0.25 in. (2.5 cm × 0.6 cm) was mounted on the sampling plate, and lowered into the probe liquid. A 10-μL air bubble was delivered to the membrane surface and two contact angles (one on each side of the bubble) were measured and averaged. A minimum of three coupons and three air bubbles per coupon were used to calculate the contact angle for each membrane. Additional details of the membrane surface energy measurements can be found elsewhere (Brant and Childress 2002).

Contact angle results are shown in Table 1.2. Contact angle measurements can be used as an indication of the hydrophobicity of the membrane, with larger contact angles signifying a more hydrophobic membrane and smaller contact angles signifying a more hydrophilic membrane. In this study, the order of increasing hydrophobicity was NF-90, NE-90, and TS-80. In general, a more hydrophobic membrane would be expected to foul more quickly, and therefore the TS-80 membrane would be expected to foul more than the NF-90 or NE-90. However, similar to the zeta potential results, there is less than 10 percent difference between the contact angle values between NF-90 and NE-90 and between NE-90 and TS-80.

Surface Roughness Measurements

Membrane surface roughness was determined using atomic force microscopy (AFM). All AFM experiments were carried out using a Park Scientific Instruments (Sunnyvale, CA) Autoprobe CP atomic force microscope. The atomic force microscope was operated in the aqueous environment using a quartz liquid cell. Silicon nitride cantilevers with integrated pyramidal tips (VEECO Instruments, Inc., Fremont, CA) were used in tapping mode. Five scans of each membrane surface were collected to ensure that the reported roughness statistics were more representative of the surface as a whole.

Table 1.2
Membrane contact angle (pH = 5.65 and T = 25°C)

Membrane	Contact Angle (°)
NF-90	48.7
NE-90	51.5
TS-80	55.1

Key roughness statistics calculated for each membrane are summarized in Table 1.3. Average roughness (R_a) is defined as the average deviation of surface asperities (i.e., peaks and valleys) from the mean plan, or half the average peak-to-valley depth. Root-mean-squared (RMS) roughness (R_q) describes the standard deviation of surface features from the mean plane. The surface area difference (SAD) describes the increase in surface area, due to roughness, over a perfectly flat plane with the same dimensions (Hoek and Elimelech 2003). It indicates the accumulated surface area of all roughness features on the membrane sample, which includes the size and number of asperities and determines the spacing of those asperities.

The membranes in order of increasing R_a were NE-90, NF-90, and TS-80. The R_a for the TS-80 membrane was approximately 10 percent greater than for the NF-90 membrane, and the R_a for the NF-90 membrane was approximately 20 percent greater than the NE-90 membrane. For the membranes tested, the R_q values increased on the same order as the R_a and are larger in magnitude than the R_a. Although the NF-90 membrane has a greater average roughness than the NE-90 membrane, it has a much lower SAD. This indicates that the NF-90 membrane has larger surface asperities than the NE-90 membrane, but in a lower density throughout the membrane surface. However, the TS-80 membrane has larger surface asperities and a greater number of them throughout the membrane than either the NF-90 or NE-90 membranes.

Surface roughness (R_a) has been shown to correlate to membrane fouling (Vrijenhoek et al. 2001). Water permeation through an NF membrane is proportional to the thickness of the membrane active layer. The thinnest section of a rough membrane will occur in the valleys and the thickest section is at the peaks. Because of this, water convection and particle transport occur preferentially in the valleys. Particles are likely to become trapped in these valleys (Vrijenhoek et al. 2001). Also, tangential forces will not be able to shear off the particles that are lodged in the valleys. Membranes with large SAD would be expected to have a higher initial flux, because the water would pass through a larger surface area. Based on these findings, the TS-80-membrane would be expected to foul more than the NF-90 and NE-90 membranes, as well as have a greater initial flux.

GOVERNING EQUATIONS FOR PROCESS MODELING

The purpose of the modeling is to predict performance of the NF dual-staged system in terms of permeate water quality based on operating parameters (temperature, pressure) and feed salinity. Equations for the model are developed below.

Table 1.3
Membrane roughness statistics

Membrane	R_a (nm)	R_q (nm)	SAD (percent)
NF-90	57.6	74.9	13.2
NE-90	46.3	59.0	24.8
TS-80	63.3	79.4	34.6

Based on mass balance and flow conservation:

$$Q_f\,C_f = Q_b\,C_b + Q_p\,C_p \tag{1.3}$$

$$Q_f = Q_b + Q_p \tag{1.4}$$

where C is the salt concentration (g/L), Q is the flow (gpd), and subscripts f, b, and p denote feed, brine, and permeate respectively. Using equations (1.4), (1.3) becomes:

$$C_b = \frac{Q_f\,C_f - Q_p\,C_p}{Q_f - Q_p} \tag{1.5}$$

Let Y be the recovery and R the salt rejection

$$Y = \frac{Q_p}{Q_f} \tag{1.6}$$

and

$$R = 1 - \frac{C_p}{C_f} \tag{1.7}$$

Rearranging equations (1.5) with (1.6) yields:

$$C_b = \frac{1}{1-Y}\left(C_f - Y\,C_p\right) \tag{1.8}$$

Using equation (1.7), equation (1.8) becomes:

$$\frac{C_b}{C_f} = \frac{1}{1-Y}\left(1 - Y\left(1-R\right)\right) \tag{1.9}$$

$$\frac{C_b}{C_f} = 1 + \frac{Y\,R}{1-Y} \tag{1.10}$$

Rather than using the thin-film theory model which involves the determination of a mass transfer coefficient that depends on hydraulic characteristics specific to the situation the membrane system is used (e.g., the Sherwood number would change at bench- or pilot- or full-scale), the research team decided to adopt the simpler and more robust solution diffusion model derived from Kedem and Katchalsky (1958). The classic equations governing water flux, J (gfd), and flux of salt i, J_i (gfd), through an RO or NF membrane are (Mulder 1996):

$$J = k_w \left(P - \sigma\,\pi\right) \tag{1.11}$$

$$J_i = k_i \left(C_{f_i} - C_{p_i}\right) \tag{1.12}$$

where P is the applied pressure (psi), π is the osmotic pressure in the bulk (psi), σ is the membrane reflection coefficient (an indication of membrane selectivity, $0< \sigma <1$), and k_w (gfd/psi) and k_i (gfd) are the temperature-dependent water and salt permeability coefficients.

Equations (1.11) and (1.12) and can also be written as:

$$J = \frac{Q_p}{A} \tag{1.13}$$

$$J_i = J\,C_{p_i} \tag{1.14}$$

where A is the membrane area (ft^2). Adapting equation (1.7) to salt i and integrating it to equation (1.12) yields:

$$R_i = 1 - \frac{C_{p_i}}{C_{f_i}} \tag{1.15}$$

$$J_i = k_i \left(C_{f_i} - C_{p_i}\right) = k_i\,R_i\,C_{f_i} \tag{1.16}$$

$$R_i = \frac{J_i}{k_i\,C_{f_i}} \tag{1.17}$$

Using equation (1.14) to express J_i in equation (1.17) yields:

$$R_i = \frac{C_{p_i}\,J}{k_i\,C_{f_i}} \tag{1.18}$$

Using equation (1.15) into equation (1.18) yields:

$$R_i = \left(1 - R_i\right)\frac{J}{k_i} \tag{1.19}$$

so that

$$R_i = \frac{J}{J + k_i} \tag{1.20}$$

Adapting equation (1.10) to salt i and using equation (1.20) yields:

$$\frac{C_{b_i}}{C_{f_i}} = 1 + \frac{Y\left(\frac{J}{J + k_i}\right)}{1 - Y} \tag{1.21}$$

Using equations (1.6) and (1.13) to express Y and J, respectively in equation (1.21) yields:

$$\frac{C_{b_i}}{C_{f_i}} = 1 + \frac{\frac{Q_p}{Q_f}\left(\frac{\frac{Q_p}{A}}{\frac{Q_p}{A} + k_i}\right)}{1 - \frac{Q_p}{Q_f}} \tag{1.22}$$

$$\frac{C_{b_i}}{C_{f_i}} = 1 + \frac{Q_p\left(\frac{1}{1 + k_i \frac{A}{Q_p}}\right)}{Q_f - Q_p} \tag{1.23}$$

$$\frac{C_{b_i}}{C_{f_i}} = 1 + \frac{1}{\left(\frac{Q_f}{Q_p} - 1\right)\left(1 + k_i \frac{A}{Q_p}\right)} \tag{1.24}$$

Using equations (1.13) and (1.11) to replace Q_p in equation (1.24) yields:

$$\frac{C_{b_i}}{C_{f_i}} = 1 + \frac{1}{\left(\frac{Q_f}{A\,k_w\,(P - \sigma\,\pi)} - 1\right)\left(1 + \frac{k_i}{k_w\,(P - \sigma\,\pi)}\right)} \tag{1.25}$$

$$\frac{C_{b_i}}{C_{f_i}} = \frac{k_w\,(P - \sigma\,\pi)\left(\frac{Q_f}{A} - k_i\right) + \frac{Q_f\,k_i}{A}}{\left(\frac{Q_f}{A} - k_w\,(P - \sigma\,\pi)\right)\left(k_w\,(P - \sigma\,\pi) + k_i\right)} \tag{1.26}$$

The van't Hoff equation that relates π to the concentration in the bulk, is (Mulder 1996):

$$\pi = \mathrm{R_g}\, T \sum_i n_i \frac{C_{f_i}}{M_i} \tag{1.27}$$

where $\mathrm{R_g}$ is the gas constant (1.206 psi.L / mol.°K), T is the temperature (°K), n is the number of dissociating ions, M is the molecular weight (g/mol). Using equation (1.27) in equation (1.26):

$$\frac{C_{b_i}}{C_{f_i}} = \frac{k_w\left(P - \sigma\,\mathrm{R_g} T \sum_i n_i \frac{C_{f_i}}{M_i}\right)\left(\frac{Q_f}{A} - k_i\right) + \frac{Q_f\, k_i}{A}}{\left(\frac{Q_f}{A} - k_w\left(P - \sigma\,\mathrm{R_g} T \sum_i n_i \frac{C_{f_i}}{M_i}\right)\right)\left(k_w\left(P - \sigma\,\mathrm{R_g} T \sum_i n_i \frac{C_{f_i}}{M_i}\right) + k_i\right)} \tag{1.28}$$

The concentration of salt i in the permeate can be obtained by rearranging equation (1.15) and using equation (1.20) for R_i:

$$\frac{C_{p_i}}{C_{f_i}} = 1 - R_i \tag{1.29}$$

$$\frac{C_{p_i}}{C_{f_i}} = 1 - \frac{J}{J + k_i} = \frac{k_i}{J + k_i} \tag{1.30}$$

Expressing J in equations (1.30) with equations (1.11) and (1.27) yields:

$$\frac{C_{p_i}}{C_{f_i}} = \frac{k_i}{k_w\left(P - \sigma\,\mathrm{R_g}\, T \sum_i n_i \frac{C_{f_i}}{M_i}\right) + k_i} \tag{1.31}$$

Equation (1.6) can be re-written using equations (1.13), (1.11), and (1.27):

$$Q_p = A\, k_w\left(P - \sigma\,\mathrm{R_g}\, T \sum_i n_i \frac{C_{f_i}}{M_i}\right) \tag{1.32}$$

and

$$Y = \frac{A\, k_w\left(P - \sigma\,\mathrm{R_g}\, T \sum_i n_i \frac{C_{f_i}}{M_i}\right)}{Q_f} \tag{1.33}$$

Equations (1.28), (1.31), and (1.33) signify that recovery and the concentration of salt i in the brine and in the permeate vary with temperature and pressure. The model was applied in increments represented by each membrane element in a vessel. Figure 1.5 describes the model steps. The membrane characteristics (σ, k_i and k_w) used in the model were derived from the benchscale tests at conditions where feed concentrations approach those of seawater. The remaining parameters are known (i.e., feed flow, membrane area, feed ionic composition, and gas constant).

1. Calculate osmotic pressure based on feed concentration (Eq. 1.27)
2. Calculate permeate flow produced by each membrane element (Eq. 1.32)
3. Calculate ion concentration in permeate for each membrane element (Eq. 1.31)
4. Calculate ion concentration in brine leaving each membrane element (Eq. 1.28)
5. Repeat steps 1-4 for next membrane element using concentrate from preceding membrane element as feed for next element (e.g., subtract permeate flow from feed flow in previous element to obtain feed flow for new element)
6. Calculate total permeate flow after calculations are done for each membrane element of Stage 1
7. Determine how many membrane elements (starting from the last one) are contributing to the amount of permeate flow being returned
8. Calculate ion concentration in the permeate flow being returned using the ion concentration in the individual permeate from the contributing membrane elements
9. Calculate flow and each ion concentration for the portion of Stage 1 permeate being fed in Stage 2
10. Repeat steps 1-5 for Stage 2
11. Calculate total permeate flow
12. Calculate weighted average of each ion concentration in Stage 2 permeate
13. Calculate total flow of Stage 2 concentrate (i.e., subtract total permeate flow from feed flow)
14. Start next loop by calculating new feed flow and ion concentrations resulting from Stage 2 concentrate and Stage 1 permeate portion being returned
15. Repeat steps 1-14 until flow and ion concentrations reach steady state.

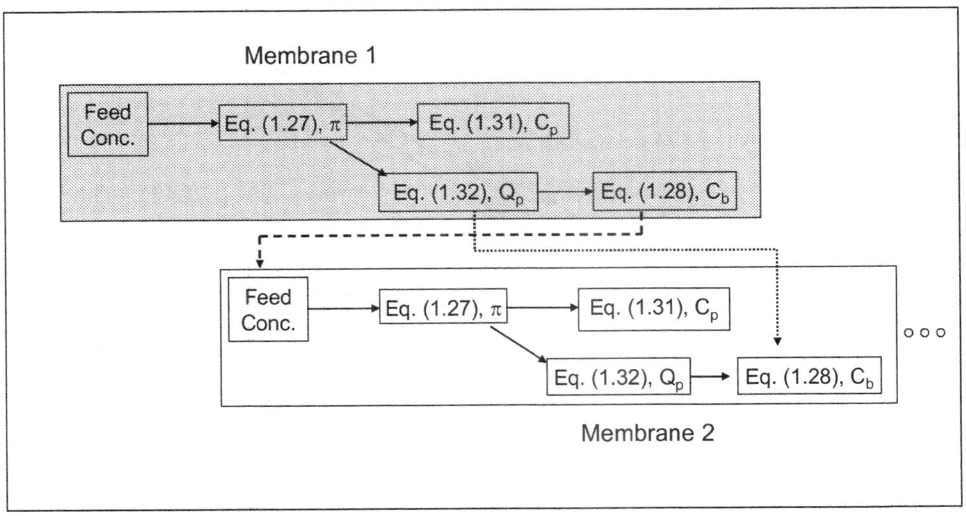

Figure 1.5 Calculation steps for predictive model

MEMBRANE PERFORMANCE

A systematic investigation of the water flux and salt rejection of the NF membranes was performed using stirred-cell and bench-scale membrane test units. Together with the membrane surface characteristics, these results assist with the selection protocol. They also provide the information necessary to model process performance at the pilot scale for the operation optimization phase.

Stirred Cell Experiments

Membrane performance characteristics were determined under batch conditions using a stainless-steel stirred cell (Sterlitech Model HP4750, Kent, WA). The stirred cell has a processing capacity of 300 mL with an active membrane area of 14.6 cm^2. It is capable of operating up to a maximum pressure of 1,000 psi. The stirred cell assembly consists of three primary components: a cylindrical body with removable top and bottom, a Teflon coated stir bar assembly to provide mixing, and a porous stainless steel membrane support disk.

Prior to each experiment the membrane was compacted at 150 psi for 4 hrs using DDW. After 4 hrs the feed pressure was reduced to 40 psi and the system was operated at this feed pressure for 1 hr to achieve flux stabilization. Permeate flux, J, was measured continuously using an analytical balance (Model PB303-3, Mettler Toledo, Columbus, OH) interfaced with a computer. The feed pressure was then increased in increments of 30 psi up to 250 psi. The DDW flux versus pressure for the membranes is shown in Figure 1.6. The flux increases linearly with pressure for all membranes. The NF-90 and NE-90 had similar DDW fluxes over the pressure range tested, while the flux for the TS-80 membrane was about half of that obtained for the NF-90 and NE-90 membranes.

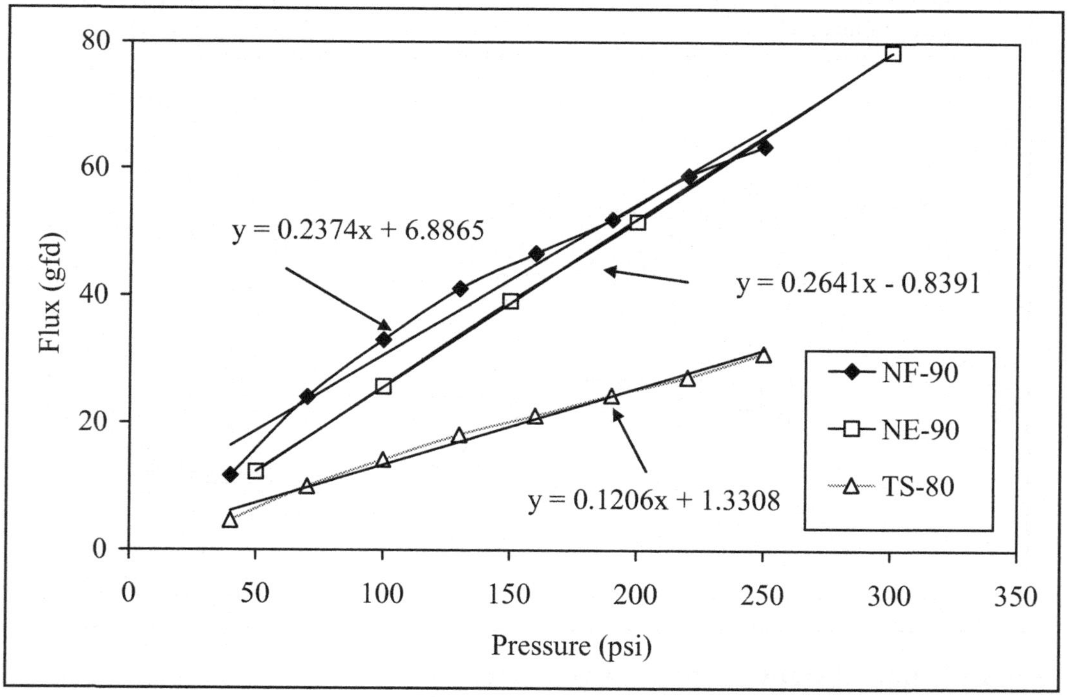

Figure 1.6 DDW water flux vs. pressure

From each plot, the water mass transfer coefficient (k_w) was calculated by fitting a linear trendline to the data and determining the slope of that line. Mass transfer coefficients are shown in Table 1.4. The NF-90 and NE-90 membranes have similar mass transfer coefficients and the TS-80 has a much lower mass transfer coefficient. As expected, the k_w values for all three nanofiltration membranes are higher than those for typical SWRO membranes. For example, Mulder (1996) gives a range of 0.002 to 0.122 gfd/psi for RO membranes and Dow Filmtec lists their SWRO membrane as having a coefficient of 0.03 gfd/psi.

At this point in the investigation, it was decided to continue the membrane performance tests with only the NF-90 and NE-90 membranes. Based on the low DDW mass transfer coefficient value for the TS-80 membrane as well as results obtained with the pilot plant, the TS-80 membrane had comparatively low flux and rejection. However, it was decided to evaluate the fouling potential of the TS-80 membrane to qualitatively correlate surface characterization and fouling results. Thus, the TS-80 membrane will be discussed in the membrane fouling section.

Bench-Scale Experiments

The next phase of the membrane performance evaluation was to determine at bench scale the water and salt permeability coefficients for each membrane in the presence of a variety of salt solutions. It was important to obtain individual salt mass transfer coefficients for each ion, because contrary to SWRO membranes that provide high rejection of all ions, NF membranes reject ions of various valence and size differently.

Bench-Scale System

The water flux and solute rejection of the membranes were evaluated using a bench-scale membrane test unit (Figure 1.7), in which the test solution was pumped from a reservoir, through a flat sheet membrane cell (model SEPA-CF, Osmonics, Minnetonka, MN), and back to the reservoir. The membrane cell was fitted with both feed and permeate spacers to simulate the hydrodynamics of a spiral-wound membrane element. The reservoir housed a stainless steel cooling coil connected to a recirculating chiller to maintain the test temperatures.

Prior to each set of experiments, the membrane was compacted at 575 psi (the highest experimental operating pressure) for 8 hr using DDW. After compaction, the salt solution was added to the feed reservoir and recirculated through the system at low pressure for the equivalent of two tank volumes to provide adequate mixing. During this time, the temperature was equilibrated to 15°C. The pressure was then increased to the initial operating pressure, and the system was allowed to reach equilibrium over 30 min, which was sufficient to stabilize the flux.

Table 1.4
DDW mass transfer coefficients

Membrane	k_w (gfd/psi)
NF-90	0.24
TS-80	0.12
NE-90	0.26

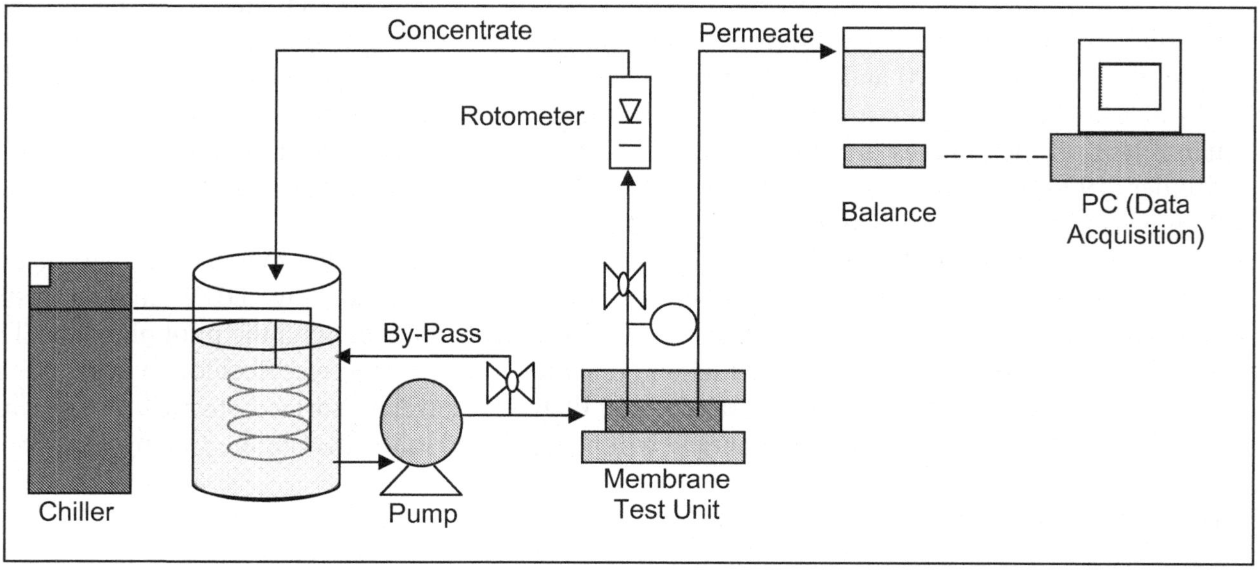

Figure 1.7 Schematic of the bench-scale membrane test system

The permeate flux, *J*, was then measured continuously for 20 min using an analytical balance (Model PB 303-3, Mettler Toledo, Columbus, OH) interfaced with a computer. The permeate was returned to the feed tank after each experiment. At the end of the 20 min experiment, two 100-mL samples were taken for analysis. This process was repeated for the remainder of the operating pressures. The temperature was then raised to 20°C, and the entire set of experiments was repeated. Between experiments, the membrane system was emptied of the salt solution and rinsed. First, the system was flushed with deionized (DI) water, and rinsed (in recirculation mode) with DDW; then, the new salt solution was added and recirculated through the system. The first permeate volumes were discarded.

Bench-Scale Test Plan

The test plan was developed to be a systematic approach to determine mass transfer coefficients for the major salts constituting the synthetic seawater. Feed concentrations were selected such that they would bracket the expected salt concentration range in the feed of Stage 1 and Stage 2 for the pilot plant. Operating pressures and temperatures were similar to those tested in the pilot plant tests. The bench-scale test plan consisted of:

- Measuring the permeate flux and ion concentrations for feedwater made of DDW and 0.07, 0.28, 0.42, and 0.56 M of NaCl at 15 and 20°C. Operating pressures for the first NaCl concentration were: 100, 150, 200, and 300 psi; for the second NaCl concentration: 250, 300, and 350 psi; for the third NaCl concentration: 350, 400, and 450 psi; and for the last NaCl concentration: 450, 500, and 575 psi
- Repeating the previous step at 15 and 20°C for each of the following salts:
  - $MgCl_2$ at 0.003 M (145, 225, 300 psi) and 0.029 M (180, 500, 575 psi)
  - $MgSO_4$ at 0.003 M (150, 225, 300 psi) and 0.031 M (170, 500, 575 psi)
  - $CaCl_2$ at 0.001 M (150, 225, 300 psi) and 0.013 M (165, 500, 575 psi)
  - NaBr at 0.001 M (150, 500, 575 psi)

- $NaHCO_3$ at 0.003 M (150, 500, 575 psi)
- KCl at 0.01 M (150, 500, 575 psi)
- KI at 6.0×10^{-7} M (150, 300, 575 psi)
- Boron at 4.6×10^{-4}M (150, 300, 575 psi; at pH 7.8, 9.8)

- Measuring the permeate at 500 and 575 psi for feedwater made of synthetic seawater (NaCl = 0.486 M, $MgCl_2$ = 0.029 M, $MgSO_4$ = 0.031 M, $CaCl_2$ = 0.013 M, KCl = 0.010 M, $NaHCO_3$ = 0.003 M, and NaBr = 0.001 M) at both 15 and 20°C. The ionic composition of the permeate was analyzed (Note: this step corresponded to Stage 1 of the dual-staged NF pilot system)
- Measuring the permeate at 200 psi for feedwater made up of the 500 psi permeate analyzed in the previous step. The ionic composition of the permeate and the conductivity were analyzed
- Measuring the permeate at 200, 300, and 400 psi for feedwater made up of the permeate obtained for the synthetic seawater at 575 psi. The ionic composition of the permeate and the conductivity were measured (Note: this step corresponded to Stage 2 of the dual-staged NF pilot system)
- Measuring the permeate at 500 psi for feedwater made up of 75 percent synthetic seawater and 25 percent Stage 1 permeate obtained at 575 psi to assess the impact of feed dilution through partial Stage 1 permeate recycle. Conductivity was analyzed to determine TDS rejection. This permeate was treated at 200 psi in a "Stage 2" test, and again the conductivity was analyzed
- Calculating the specific flux and salt passage for all of the above conditions to better match the parameters used in existing seawater desalination plants and studies from the Consorcio de Fuereventa SWRO plant in the Canary Islands (Moss and Gil 1999) and the Doha SWRO Research Plant in Kuwait (Abdel-Jawad et al. 2001). These parameters are reported in Appendix A.

Parameter Analysis

Ions were analyzed using several methods. Sodium and potassium were measured using flame emission photometry (Standard methods 3500-Na B and 3500-K B); calcium and magnesium using flame atomic absorption spectrometry (AAS) (Standard method 3111 B); boron using inductively coupled plasma ICP spectrophotometry (Standard method 3120 B); sulfate, bromide, iodide and chloride with ion chromatography (IC) (Standard method 4110 A); and bicarbonate using an alkalinity titration (Standard method 2320 B).

k_w was calculated using the same method as described in the stirred-cell section. For each solute i, k_i was calculated by rearranging Eq. (1.19) and σ_i by using Eq. (1.34) below, which is based on Mulder (1996). The k_i and σ_i values obtained at specific salt concentrations and temperatures were then averaged over the flux (i.e., pressure).

$$\sigma_i = 1 + \frac{(1-R_i)\ln(1-R_i)}{R_i} \tag{1.34}$$

Single Salt Results

Flux results for the NF-90 membrane are shown in Figure 1.8 for the NaCl feed solution. Flux increased linearly with pressure simply because of the increased driving force. Flux increased with increasing temperature due to the decreased viscosity of the water. Flux also increased with decreasing feed concentration due to the decreased osmotic pressure. Results for the other salts tested as well as for the NE-90 membrane followed similar trends.

Some of the permeate flux versus applied pressure lines seen from the bench-scale experiments had a positive y-intercept. This is not reasonable in a practical sense because at an applied pressure of zero, the process cannot still be producing water. Other studies (Bowen and Mukhtar 1996 and Turan 2004) report flux versus pressure for several NF membranes, and some of their data also shows a positive y-intercept. An explanation of the slope behavior was not provided, but several hypotheses, either individually or in combination, could explain these results:

- The membranes were not compacted completely, resulting in an observed flux higher than it should be. In the bench-scale experiments however, the flux stayed constant at each of the experimental pressures, suggesting that the membranes were fully compacted before the experiments began

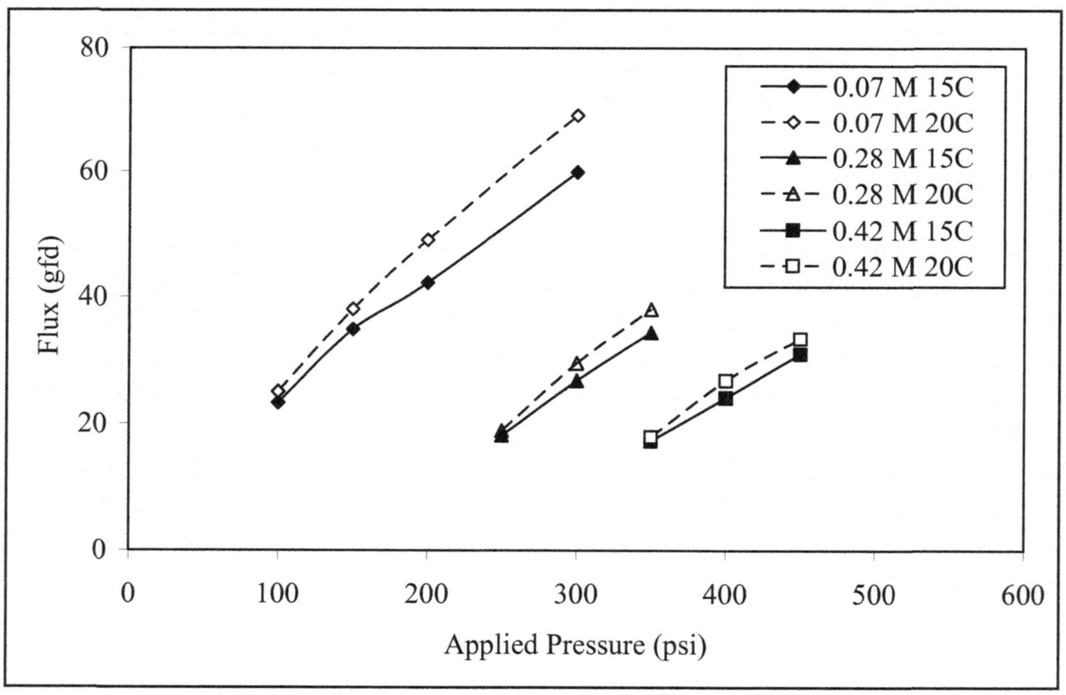

Figure 1.8 Permeate water flux for the NF-90 membrane experiments with NaCl feed solution (I = 0.07 M, 0.28 M, 0.42 M NaCl and T = 15°C and 20°C)

- Although membranes were compacted at the highest pressure run in the experiments, once this pressure was released, the membranes may have relaxed slightly and consequently the flux may have increased above what it should have been at low pressures. This could slightly change the slope of the lines, which can have large impacts on the y-intercepts. The slopes of experimental lines were similar whether the stirred-cell, SEPA cell, or smaller black membrane cells were used, confirming that the phenomena exists on all the bench-scale setups used for this project. Because of the complex nature of compaction, its effects should be investigated further in the future
- A systematic error in the gauges undervalued actual pressures. The pressure gauges used in the bench-scale experiments had a ± 10 psi error. For the NE-90 membrane, an error of 10 to 30 psi was generally needed to adjust from a positive to a negative y-intercept. Also, less adjustment of the 20°C line was needed than to the 15°C line because of the higher slope seen at 20°C. For the NF-90 membrane, an error of 5 to 15 psi was generally needed to adjust the y-intercept to a negative value. Again, less adjustment was needed at 20°C. If the error in the gauges was similar across the entire range of pressures, it would not affect the slope of the lines but rather only shift the lines horizontally
- The point at the highest pressure tested (575 psi) on several sets of experiments exhibited a leveling off, which may mean that the limiting flux is being approached. This point caused the slope of the line to decrease and sometimes pushed the y-intercept positive. Where this was observed, removing this point often changed the slope enough to make the intercept negative.

Sodium rejection results for the NF-90 membrane are shown in Figure 1.9 for the NaCl feed solution. Rejection generally increased with pressure because more water flows through the membrane while salt transport remains about same. However, after a critical pressure was exceeded for some salts, rejection began to decrease.

A similar trend was observed at pilot scale and was referred to as the "critical point" (Tseng et al. 2005). Below the critical point, convection dominates and permeate water quality improves with increasing pressure. Once the critical point is reached, diffusion dominates and permeate water quality deteriorates with further increases in pressure (Tseng et al. 2005). The critical pressure for NaBr is shown in Figure 1.10, which shows bromide rejection increasing with applied pressure until approximately 500 psi. At this point, the rejection begins decreasing with increasing pressure. In bench-scale experiments, only three pressures were tested at each salt concentration and so if this phenomenon were to be verified at the bench-scale, more points would need to be included.

Rejection was not greatly affected by water temperature at the conditions tested. Rejection decreased with increasing feed concentration due to an increased concentration at the membrane surface, which allowed more salt to diffuse through the membrane. Trends for the NE-90 membrane were similar to the NF-90, but rejection levels were generally lower.

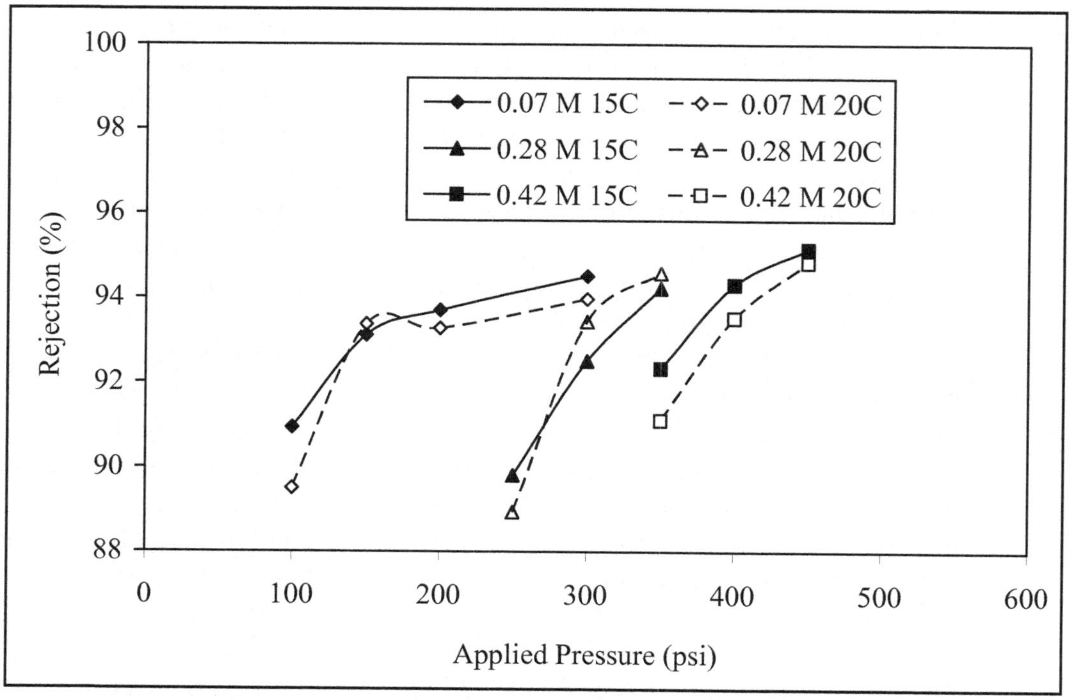

Figure 1.9 Sodium rejection for the NF-90 membrane experiments with NaCl feed solution (I = 0.07 M, 0.28 M, 0.42 M NaCl, and T = 15°C and 20°C)

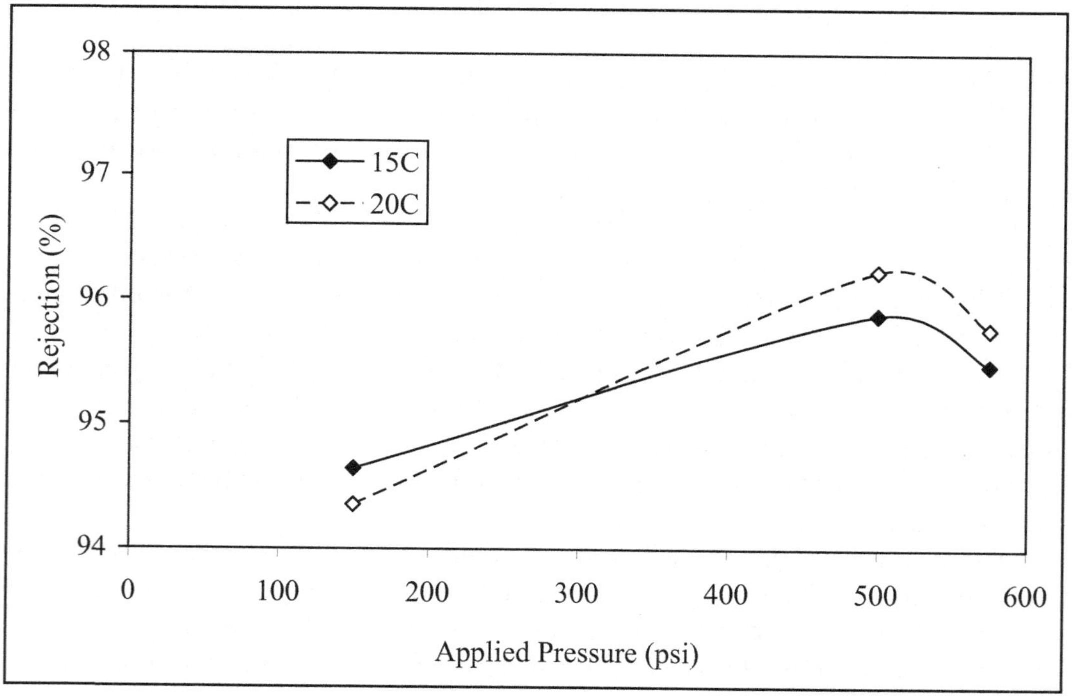

Figure 1.10 Bromide rejection for the NF-90 membrane experiments with NaBr feed solution(I = 0.001M NaBr and T = 15°C and 20°C)

A summary of the operating conditions and the values obtained for k_w, k_i, and σ_i values are shown in Tables 1.5 and 1.6 for NF-90 and NE-90, respectively. The flux and rejection data for the single salt experiments for NF-90 are included in Appendix A. Several general trends were observed from the data. The k_w decreased with increasing salt concentration. This is most likely due to concentration polarization. It is generally accepted that k_w is constant for RO membranes (Mulder 1996). However, because NF membranes have a lower rejection than RO membranes, the CP layer thickness may be more influenced by feed concentration. The dependency of k_w on feed water concentration was also observed in the pilot plant. Further, increasing temperature increases k_w because the water viscosity decreases.

Increasing temperature also increased k_i. For boron, k_i greatly decreased with increasing pH due to the transition in speciation from boric acid to borate at higher pH values. Whereas the uncharged boric acid is able to pass through the membrane, the charged borate ion is better rejected. The k_i for monovalent ions was found to be higher than for divalent ions, which is expected for NF membranes. Also, the k_i for bicarbonate is lower than for the rest of the monovalent ions most likely due to its larger size. Temperature had overall little effect on the reflection coefficient for both membranes. At a given concentration of NaCl, $CaCl_2$, and $MgCl_2$ the difference in reflection coefficients between cations and anions is more pronounced for NF-90 than for NE-90. Overall, NF-90's σ values were slightly higher than NE-90's, indicating higher membrane selectivity towards the ions in solution. The reflection coefficient behavior towards boron is strongly dependent on solution pH (as it was for k_i).

Synthetic Seawater Results

Table 1.7 shows k_w and k_i for the synthetic seawater experiments that simulated Stage 1 of the pilot plant. The NF-90 and NE-90 membranes had similar k_w values, approximately 0.1 gfd/psi. This k_w value matched well with the k_w seen for the single salt NaCl experiment of similar ionic strength to the seawater. Both membranes also had k_i values for divalent ions that were substantially less than for monovalent ions. The NF-90 membrane had lower k_i values than the NE-90 membrane for all ions.

The membranes' ability to remove monovalent ions is a good indication of their ability to desalinate seawater. The nature of NF membranes allows them to remove high levels of divalent ions. However, the main component in seawater is NaCl making single salt rejection a very important aspect in selecting an applicable NF membrane for seawater desalination. Also, when estimating the k_w for seawater solutions, a feedwater made up of NaCl with a similar ionic strength was found to be a suitable surrogate.

For both membranes, the mass transfer coefficient value for the overall salinity, k_{TDS}, which was determined based on conductivity of the feed and permeate, is remarkably close to the value obtained from combining the individual k_i on a mass basis (the molar basis did not provide as good a match). Overall, the mass transfer coefficients were insensitive to the change in pressure in Stage 1 for both membranes.

Table 1.8 shows k_i and k_w for the simulated Stage 2 experiments, which used Stage 1 permeate obtained with two different pressures (500 and 575 psi). Stage 2 k_w values are higher than for Stage 1 because feed concentration (and hence the CP layer) is smaller. The k_w value for each membrane remains same whether feed was composed of Stage 1 permeate at 575 psi or at 500 psi. Also, k_w for NF-90 had a much larger increase than for NE-90 membrane from Stage 1 to Stage 2 due to its higher ion rejection. The k_i values for all ions were lower in Stage 2 than

Stage 1, again due to lower feed concentrations. As in Stage 1, the k_i values for divalent ions are substantially lower than for monovalent ions.

The individual salt mass transfer coefficients for NF-90 remained unaffected by the pressure in Stage 2. Changing the feed composition by using Stage 1 permeate obtained at 500 psi instead of Stage 1 permeate obtained at 575 psi reduces by about half the k_i values. On the other hand, NE-90 appears to be less sensitive to feedwater makeup but more susceptible to Stage 2 pressure, with a 25 to 40 percent increase in k_i values for most solutes when Stage 2 pressure is increased from 200 to 300 psi. For ions with comparatively large atomic mass units (i.e., bromide and sulfate), this behavior is not observed.

Table 1.5
NF-90 bench-scale operating conditions and mass transfer coefficients

Salt		Feed Conc. (M)	Temp.(°C)	k_w (gfd/psi)	k_i (gfd) cation	σ_i cation (percent)	k_i (gfd) anion	σ_i anion (percent)
NaCl		0.07	15	0.18	2.81	80	2.09	84
			20	0.22	3.41	79	2.16	84
		0.28	15	0.16	2.11	79	1.26	85
			20	0.19	2.20	79	1.29	85
		0.42	15	0.14	1.49	82	0.96	87
			20	0.15	1.81	80	1.12	86
		0.56	15	0.10	2.14	77	1.41	83
			20	0.11	2.56	75	1.60	82
$CaCl_2$		0.001	15	0.22	0.42	96	1.30	90
			20	0.26	0.51	96	1.53	91
		0.013	15	0.19	0.16	99	1.02	94
			20	0.22	0.18	99	1.21	94
$MgCl_2$		0.003	15	0.24	0.61	95	2.22	87
			20	0.31	0.47	96	1.86	90
		0.029	15	0.19	0.31	98	0.84	95
			20	0.23	0.36	98	0.94	96
$MgSO_4$		0.003	15	0.19	0.09	99	0.10	99
			20	0.23	0.08	99	0.10	99
		0.031	15	0.14	0.17	99	0.26	98
			20	0.17	0.18	99	0.29	98
KCl		0.01	15	0.19	4.87	81	4.76	81
			20	0.22	5.06	82	4.67	83
NaBr		0.001	15	0.19	3.90	83	3.63	85
			20	0.21	4.10	85	3.79	85
$NaHCO_3$		0.003	15	0.16	1.17	94	0.87	95
			20	0.19	0.87	95	0.58	97
KI		6×10^{-6}	15	0.19	-	-	9.03	71
			20	0.23	-	-	10.38	71
Boron	pH 7.8	0.000463	15	0.17	30.28	41	-	-
			20	0.19	43.89	36	-	-
	pH 9.8		15	0.15	3.84	80	-	-
			20	0.15	8.32	70	-	-

Table 1.6
NE-90 bench-scale operating conditions and mass transfer coefficients

Salt		Feed Conc. (M)	Temp. (°C)	k_w (gfd/psi)	k_i (gfd) cation	σ_i (percent) cation	k_i (gfd) anion	σ_i (percent) anion
NaCl		0.07	15	0.20	1.94	78	2.55	75
			20	0.26	2.30	79	2.84	76
		0.28	15	0.14	2.12	75	2.54	72
			20	0.19	2.39	76	2.82	73
		0.42	15	0.12	2.17	75	2.66	72
			20	0.15	2.63	74	3.15	71
		0.56	15	0.10	2.40	73	2.91	70
			20	0.11	2.91	77	3.65	74
$CaCl_2$		0.001	15	0.20	2.57	84	2.85	83
			20	0.25	2.73	84	2.96	84
		0.013	15	0.18	3.79	84	3.50	85
			20	0.23	4.17	85	3.94	86
$MgCl_2$		0.003	15	0.23	2.97	82	3.31	81
			20	0.28	3.18	83	3.71	82
		0.029	15	0.17	2.48	89	2.77	88
			20	0.21	2.18	90	2.94	88
$MgSO_4$		0.003	15	0.22	2.02	87	2.50	85
			20	0.27	1.77	89	2.00	88
		0.031	15	0.18	1.95	91	2.40	89
			20	0.21	2.21	90	2.65	89
KCl		0.01	15	0.16	3.29	85	4.20	82
			20	0.20	3.60	85	4.75	82
NaBr		0.001	15	0.18	4.64	78	3.05	87
			20	0.23	3.58	86	3.94	85
$NaHCO_3$		0.003	15	0.17	1.24	93	0.49	96
			20	0.21	1.60	92	0.83	95
KI		6×10^{-6}	15	0.17	-	-	6.24	75
			20	0.20	-	-	7.10	76
Boron	pH 7.8	0.000463	15	0.18	20.9	54	-	-
			20	0.23	32.3	48	-	-
	pH 9.8		15	0.21	10.0	70	-	-
			20	0.25	16.8	64	-	-

Table 1.7
Mass transfer coefficients from synthetic seawater at 20°C (simulated Stage 1)

	k_w (gfd/psi)	P (psi)	k_{Na+} (gfd)	k_{K+} (gfd)	k_{Ca2+} (gfd)	k_{Mg2+} (gfd)	k_{Cl-} (gfd)	k_{Br-} (gfd)	k_{HCO3-} (gfd)	$k_{SO4\,2-}$ (gfd)	k_{TDS} (gfd)
NF-90	0.088	500	1.93	2.38	0.07	0.04	1.56	2.33	0.26	0.02	1.34
		575	2.00	2.27	0.07	0.04	1.52	2.19	0.27	0.03	1.32
NE-90	0.077	500	3.69	5.34	1.72	1.32	3.68	4.26	2.02	1.11	3.34
		575	3.80	5.55	1.82	1.41	3.88	5.39	2.11	1.02	3.56

Table 1.8
Mass transfer coefficients from simulated Stage 2 experiments using Stage 1 permeate obtained with two different pressures (T = 20 °C)

Membrane	P1 (psi)	k_w (gfd/psi)	P2 (psi)	k_{Na+} (gfd)	k_{K+} (gfd)	k_{Ca2+} (gfd)	k_{Mg2+} (gfd)	k_{Cl-} (gfd)	k_{Br-} (gfd)	k_{HCO3-} (gfd)	$k_{SO4\ 2-}$ (gfd)
NF-90	500	0.239	200	2.46	2.08	ND	ND	2.08	3.84	ND	ND
			300	2.62	2.05	ND	ND	2.05	4.30	ND	ND
	575	0.242	200	1.45	1.19	ND	ND	1.35	2.39	ND	ND
			300	1.51	1.20	ND	ND	1.43	2.60	ND	ND
NE-90	500i	0.166	200	2.11	2.42	1.25	1.06	2.30	2.00	1.41	0.38
			300	2.71	3.09	1.73	1.35	3.13	2.10	1.86	0.39
	575	0.164	200	1.93	1.92	1.20	0.86	2.29	1.76	1.29	0.32
			300	2.49	2.91	1.67	1.23	2.98	2.18	1.75	0.44

Note: ND indicates that the levels of ions in Stage 2 permeate were not detectable by the methods used. Because the feed concentrations were also very low, resulting k_i's were out of range and thus were not reported
P1: pressure with which simulated Stage 1 permeate was obtained (permeate is used for simulated Stage 2 tests)
P2: pressure for the simulated Stage 2 experiments

These observations could mean that the NE-90 uses essentially size exclusion when rejecting ions (i.e., beyond a certain size, removal level remains somewhat unchanged). In turn, this would mean that NF-90 uses electrostatic rejection to a greater degree than NE-90, which could explain the variation in reflection coefficients between anions and cations in Table 1.5 (whereas no such variations were observed for NE-90 in Table 1.6).

Table 1.9 shows the flux and TDS rejection of the NF-90 and NE-90 membranes for feed waters made up of 75 percent synthetic seawater and 25 percent permeate from Stage 1 at 575 psi from each membrane, respectively. This test simulated the effect of recycle streams (i.e., dilution of the feed stream) in Stage 1 of the pilot plant. The NF-90 membrane has a higher TDS rejection than the NE-90 membrane, not only due to the properties of the membrane (based on k_i results from previous tests) but also because of the lower overall ionic strength of the feed solution (i.e., for the same pressure more water passes through the membrane which further dilutes the permeate, resulting in apparent higher rejection). The feed ionic strength was lower in the case of the NF-90 because the salinity of the Stage 1 permeate (which was used to make the blend) obtained with NF-90 membrane was lower than with NE-90. In other words, recycling had a greater positive impact on NF-90 than on NE-90.

Table 1.9
Flux and rejection at 575 psi and 20°C for a mix of 75 percent synthetic seawater and 25 percent Stage 1 permeate obtained at 575 psi

Membrane	Flux (gfd)	Rejection TDS (percent)	k_{TDS} (gfd)	σ_{TDS} (percent)
NF-90	46.0	95.4	2.22	85
NE-90	38.7	88.8	4.88	72

Table 1.10 shows the flux and rejection of the NF-90 and NE-90 membranes for feed water made from the permeate of the experiment shown in Table 1.9 (i.e., simulating Stage 2 with the new permeate from Stage 1). Again, the NF-90 membrane had a higher flux and rejection than the NE-90 membrane. Both membranes had an increased rejection compared to Table 1.9, due to the decreased feed concentration. Flux comparison between Tables 1.9 and 1.10 cannot be made because the experiments were run at different feed concentrations and different applied pressures – the two of which have competing effects. However, the fluxes matched well with the NaCl single salt experiments at similar concentrations.

MEMBRANE FOULING AND CRITICAL FLUX

The purpose of this study was to evaluate the extent of flux decline and whether a critical flux could be determined. The experiments were run on a bench-scale system similar to that used in the membrane performance tests. However, instead of using a SEPA membrane cell, a smaller cell with a membrane surface area of 3 in^2 (19 cm^2) was used. Also, feed and permeate spacers were not used in these experiments. In order to allow for longer duration testing with constant transmembrane pressure, the system was automated by adding a control valve and a pressure transducer, both of which were controlled by LabView software.

Particle Size Analysis

To better identify the potential foulants in the feedwater and to gain greater insight into the mechanisms of flux decline, a particle size analysis was carried out on the raw and filtered (with a 5-μm cartridge filter) seawater. A ZetaPALS (Brookhaven, Holtsville, NY) instrument equipped with a 90Plus particle size analyzer was used. ZetaPALS determines particle size using dynamic light scattering, where a laser light is focused on a sample and a photomultiplier tube measures the intensity of the light that the particles scatter. This intensity is then related to the particle size. Figure 1.11 shows the particle size analyses for the raw and filtered seawater.

The raw seawater mostly consisted of particles in the 1 μm size range with a small fraction in the 100 nm range. The filtered seawater only contained the particles in the 100 nm range. Thus, from the particle analysis, it appears that the 5 μm cartridge filter was successful in removing all of the 1-μm particles. Because the solution had to be concentrated to a certain point before the particle size analysis could be performed, it was found that both the raw and filtered seawater contained very low concentrations of particles. The low concentrations increased the error in the measurement, and because of this error, the two peaks around 100 nm most likely represent the same particles.

Table 1.10
Flux and rejection at 200 psi and 20°C for permeate resulting from Table 1.9 test

Membrane	Flux (gfd)	Rejection TDS (percent)	k_{TDS} (gfd)	σ_{TDS} (percent)
NF-90	43.9	97.7	1.03	91
NE-90	32.2	94.1	2.02	82

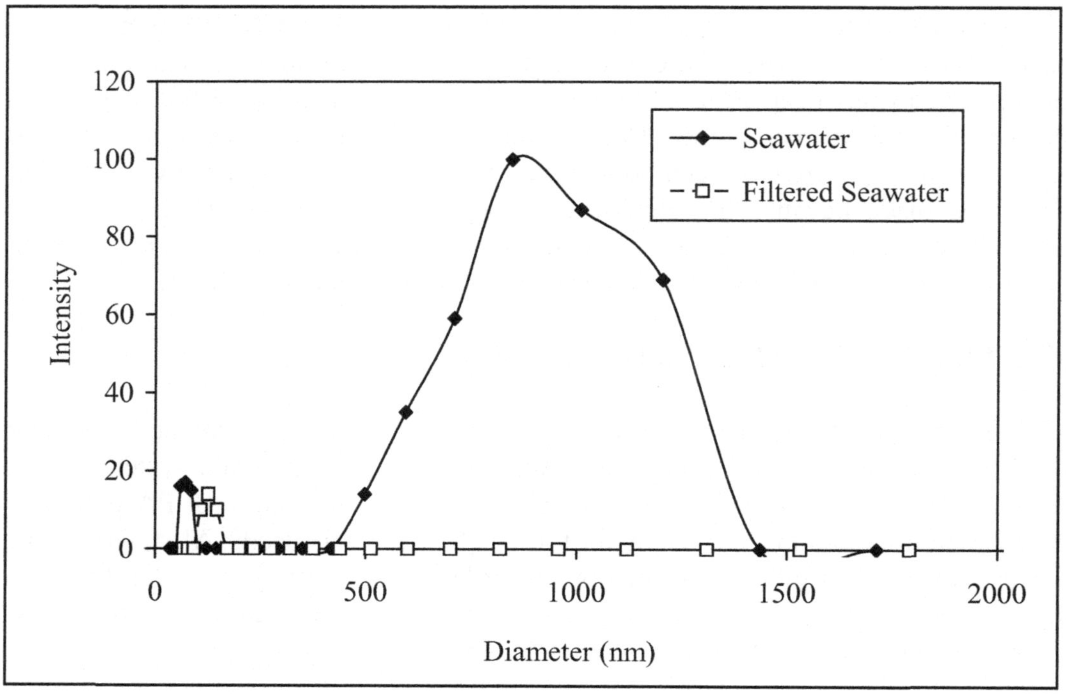

Figure 1.11 Particle size analysis for raw and filtered LBWD seawater

72-hr Fouling Experiments

The general fouling characteristics of the membranes were determined by evaluating their flux decline over a 72-hr period for a constant applied pressure and flow rate. Before each experiment, the membranes were compressed using the methods described in the membrane performance section. The feed solution, which consisted of 100 L of Pacific Ocean sea water obtained from LBWD (hereafter referred to as LBWD seawater), was then added to the feed tank. The pressure applied was 575 psi, because this was the highest pressure used in the membrane performance experiments. The flow rate of 1 L/min corresponded to the cross-flow velocity used in the membrane performance experiments. The flux decline for the three membranes is shown in Figure 1.12. The NF-90 membrane had slightly lower flux than the NE-90 membrane throughout the entire run. Both had approximately 17 percent flux reduction over the 72-hr period, and both lost flux at a similar rate. This similarity in flux decline is supported by the characterization data because both membranes had similar surface properties.

From the membrane characteristics, the TS-80 membrane was expected to exhibit greater flux decline than the NF-90 and NE-90 membranes. Also, it was seen in stirred-cell and preliminary bench-scale experiments that the TS-80 membrane had a lower initial flux. However, in the 72-hr fouling experiments, the TS-80 membrane had a similar initial flux and flux decline as the other two membranes. This indicates that membrane surface characteristics may have a less influential role on membrane fouling than the rejection properties (and in turn CP) of the membrane. Because the TS-80 membrane has a lower TDS rejection compared to the NE-90 or NF-90 membranes (approximately 82 percent compared to 89 percent or 92 percent), the CP layer will be less thick. This decreased CP layer thickness allows for a higher water flux through the membrane which, in turn, will lead to a greater fouling rate due to increased particle deposition.

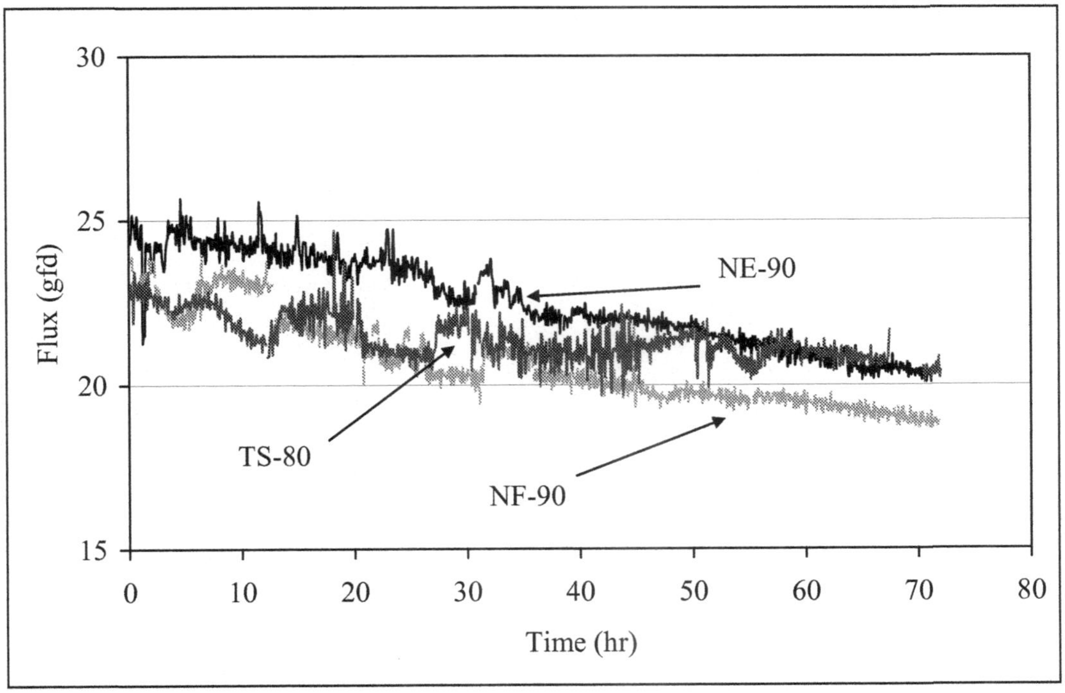

Figure 1.12 Membrane flux for the NE-90, NF-90, and TS-80 membranes (P = 575 psi, T = 15°C, Q = 1 L/min)

Critical Flux

To further evaluate flux decline and to determine if a critical flux value existed, the method proposed by Manttari and Nyström (2000) was used. First, LBWD seawater was added to the feed tank and the pressure was set to 300 psi (the lowest pressure at which permeate flux could be obtained). Permeate flux was measured continuously for 1 hr (as compared to the 15 min used in Manttari and Nyström). The pressure was then increased by 50 psi increments up to 550 psi and flux was measured for 1 hr at each pressure. The flux data is shown in Figure 1.13.

NE-90's flux increases linearly over the entire range of pressures for both the DDW and the seawater. Because the slope of the seawater flux line is less than the DDW line, it indicates that the system is operating above the strong critical flux. This suggests that some fouling phenomena are occurring but are not significantly affecting membrane performance. Because the seawater flux increased linearly with pressure, it indicates that the system has not yet exceeded the weak critical flux or that a weak critical flux could not be identified. Due to pressure limitations of the bench-scale test system, the system could not be taken above 600 psi to determine if a critical flux exists at higher pressures.

The inability to identify a critical flux in the pressure/flux range of interest was not unexpected because of the relatively low flux decline that was observed in the first few hours of the experiment (as compared to other critical flux studies where substantial fouling is seen in the first hour of testing). Because of the dilute nature of the feed water, it is likely that the flux decline observed in the 72-hr experiments is due in large part to concentration polarization and increased osmotic pressure. Although Manttari and Nyström (2000) do not differentiate between flux loss due to particle deposition and flux loss due to CP, it is likely that flux loss due to CP will not readily result in a critical flux value.

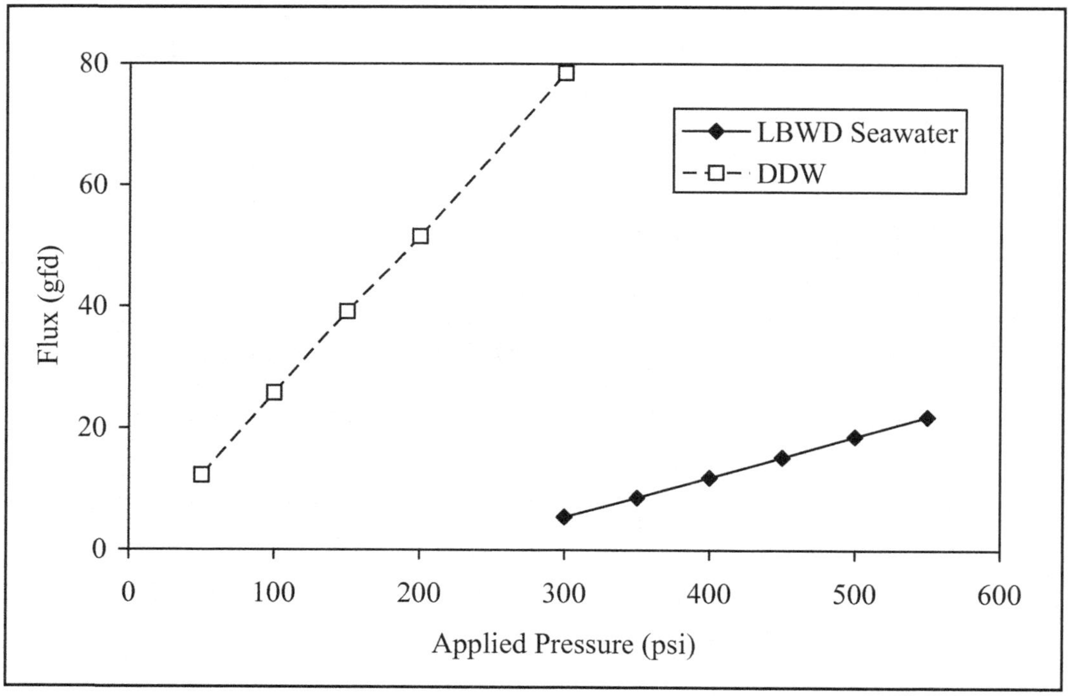

Figure 1.13 NE-90 flux *versus* pressure for DDW and LBWD seawater

Figure 1.14 shows SEM images focused on areas of concentrated deposits on the fouled NF-90 membrane surface after the 72-hr experiments. Most of the fouled membrane surface area did not contain substantial deposits. In the areas without deposits, a large amount of carbon was detected, which indicates an organic coating on the membrane surface. The TOC level of LBWD seawater is low, at approximately 0.5 mg/L (Andrews-Tate, personal communication, July 19, 2004). However, the organics that are present can adsorb to the membrane surface, decreasing water flux, and promote the growth of microorganisms on the membrane surface. Luo and Wang (2001) found that the organic foulant layer of RO desalting membranes consisted of fatty acids, carbonyl ester, and aromatic species. It was also shown that organic matter can form complex foulants with particulates, colloids (Si-Al-Fe), organics, and microorganisms.

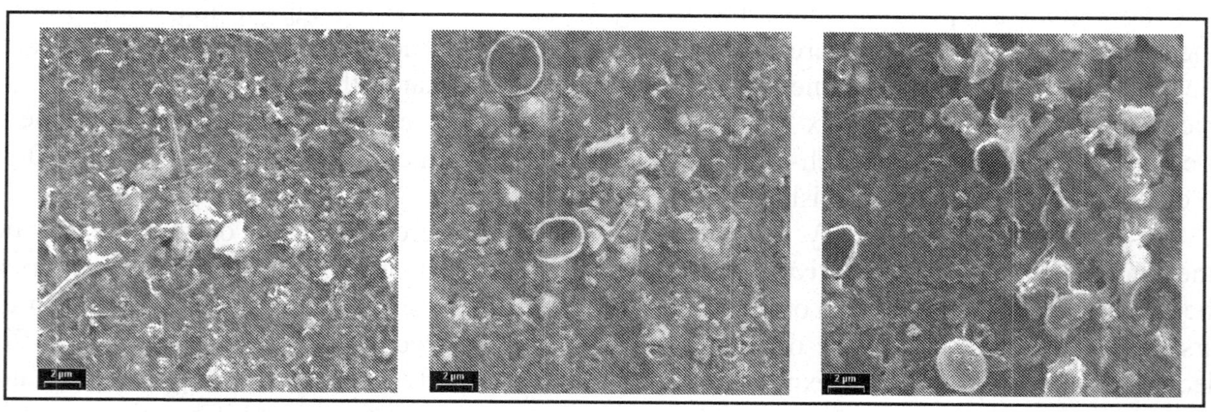

Figure 1.14 SEM images of the fouled NF-90 membrane surface at ×5000

The round structures seen in Figure 1.14 contained high levels of silica, suggesting that they are most likely diatoms. Other areas contained elements indicating that scale is present (e.g., Al, Si, and Ca). The salinity of LBWD seawater is approximately 37 g/L. Salts can cause flux loss by creating a CP layer or by scaling. To assess the sulfate scaling in the form of calcium sulfate, the solubility product (K_{sp}) of the salt was compared to its ion product (IP) in the concentrated stream. At a recovery of 30 percent, the IP for $CaSO_4$ exceeded the K_{sp}, and therefore the solution is considered to be supersaturated. Because of the probability of scaling, preventative measures such as adjusting the recovery or adding antiscalants must be used. However, scaling should be less of an issue for SWNF membranes than it is for SWRO membranes, due to their decreased rejection and CP layer thickness.

Few particles were identified in the SEM images. This is likely because the particle concentration of the LBWD seawater is very low. Also, of the diatoms and particles identified in the SEM images, some were larger than 100 nm. It is likely that these did not show up in the particle size analysis for two reasons. First, the diatoms are present at such low concentrations that they are unlikely to show in particle size analyses. Second, it is likely that some of the particulate matter aggregated during the membrane drying process that is required to take place before SEM imaging. In general, the SEM images and scaling calculations suggest that flux decline is mostly a result of CP and some scaling. The next experiments were conducted to investigate this issue further.

Concentration Polarization

The purpose of this set of experiments was to analyze the CP layer that develops during seawater desalination by NF. The two factors affecting CP are applied pressure and cross-flow velocity (Mulder 1996). The pressure was held constant at 575 psi and three flow rates were tested. Each experiment was performed on a new membrane coupon that was compacted using the methods described in the membrane performance section. The flux was measured continuously during the 72-hr run, and conductivity was measured periodically to monitor TDS rejection. The mid-range flow (1 L/min) corresponds to the velocity (194 mm/s) used in the membrane performance experiments. The two other flow rates evaluated were 0.2 L/min and 2 L/min. These correspond to cross-flow velocities of 38.9 mm/s and 389 mm/s respectively.

The flux and rejection data from these experiments are shown in Figure 1.15 and Table 1.11. It was found that the flux and rejection decreased with decreasing cross-flow velocity. This is due to the increased thickness of the CP layer. With a higher ion concentration at the membrane surface, both particle diffusion through the membrane and osmotic pressure will be increased; this will result in decreased TDS rejection and permeate flux. The flux decline for the 0.2 L/min and 1 L/min experiments was essentially the same at approximately 20 percent over the 72-hr experiment. The flux decline for the 2 L/min experiment was slightly stronger than the other two. This is probably due to the greater initial flux causing an increase in the particle transport to the surface, which the cross flow was not able to overcome.

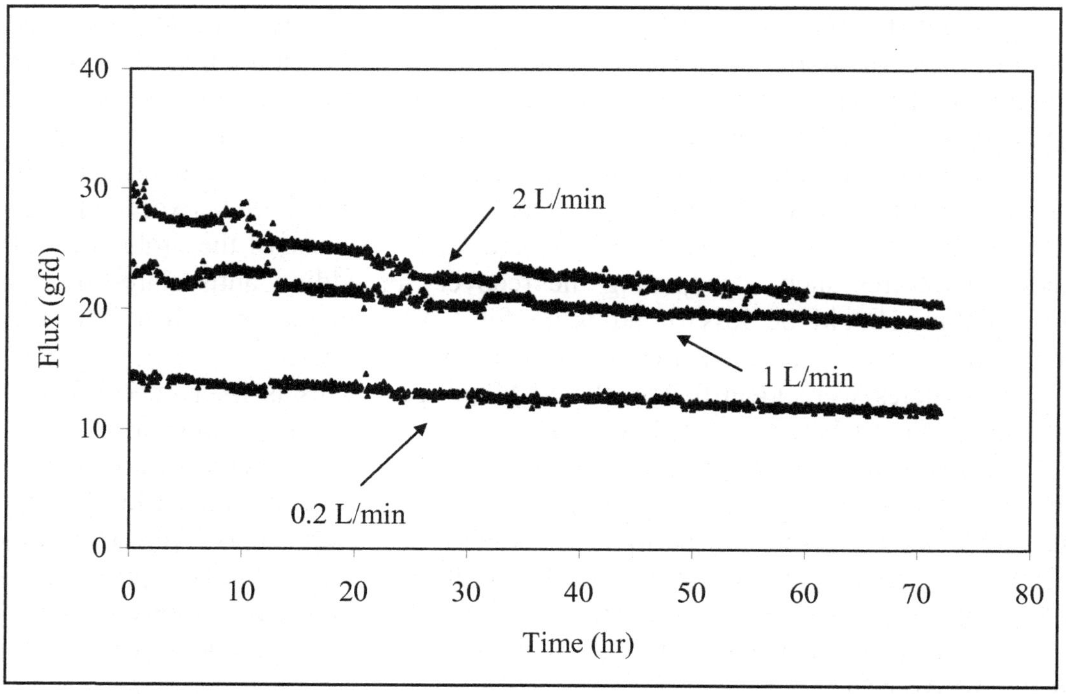

Figure 1.15 72-hr fouling tests for NF-90 membrane (P = 575 psi and T = 15°C)

Table 1.11
TDS rejection for the NF-90 membrane (P: 575 psi, T: 15°C, Q: 0.2, 1, and 2 L/min)

Feed Flowrate (L/min)	Rejection (percent)
0.2	82
1.0	90
2.0	92

Conclusions

The NF-90, NE-90, and TS-80 membranes all had similar initial flux and flux decline in the 72-hr fouling experiments. The NF-90 and NE-90 membranes had similar surface characteristics, and thus this result was expected. However, the surface roughness and hydrophobicity data of TS-80 suggested that fouling would have been more pronounced. This indicates that, in this case, membrane surface characteristics may have a less influential role on membrane fouling than rejection properties (and, in turn, CP) of the membrane.

The critical flux concept was not an effective means to evaluate the effects of operating conditions on membrane fouling for SWNF desalination. Since the critical flux concept was derived for colloidal and particulate foulants, it may not directly apply to dissolved foulants, which were prevalent in the seawater. Also, since fouling may occur more slowly for the dilute suspensions tested in this study, it was difficult to identify a critical flux within the experimental constraints and in a reasonable time period.

From SEM imaging and surface elemental analysis, it was determined that the majority of fouling can be attributed to a few localized deposits on the membrane surface as well as an organic coating over the entire surface. Elemental analyses revealed that the deposits consisted of diatoms (high in silica) as well as scale (e.g., Si, Fe, Al, and Ca).

Cross-flow velocity greatly affected flux and rejection of the NF-90 membrane. As cross-flow velocity increases, the flux and rejection increases due to decreased thickness of the CP layer, showing that decreasing the CP layer thickness is important for optimizing the system as a whole.

CHAPTER 2
OPERATIONAL OPTIMIZATION

BACKGROUND

The optimization phase used LBWD's existing 8-gpm dual-staged NF pilot plant. Two commercially-available NF membranes (Dow Chemical FilmTec NF-90 and Saehan NE-90) were used in 100 pilot tests. The feed pump was fitted with a heat exchanger to maintain a constant temperature during each test. Although the seawater salinity remained fairly constant at about 34,500 mg/L, the temperature varied through the year, and was thus expected to affect the process performance. Therefore, the optimization included the impact of feedwater temperatures by bracketing the yearly seasonal variations from the ocean near Long Beach (15 to 20 °C).

For each selected feedwater temperature, the pressure from the first stage (high pressure) covered a 470 to 580-psi range, while the operating pressure in the second stage varied from 150 to 250 psi. The pilot plant configuration is shown in Figure 2.1. Flow streams recycled to the feed include Stage 2 concentrate (at all times) and various amounts of the more saline part of Stage 1 permeate (e.g., 0, 20, and 40 percent of the total Stage 1 permeate flow). These recycle streams do not change the fundamental nature of the process and were included in the test plan to provide refinements that would possibly further enhance overall recovery.

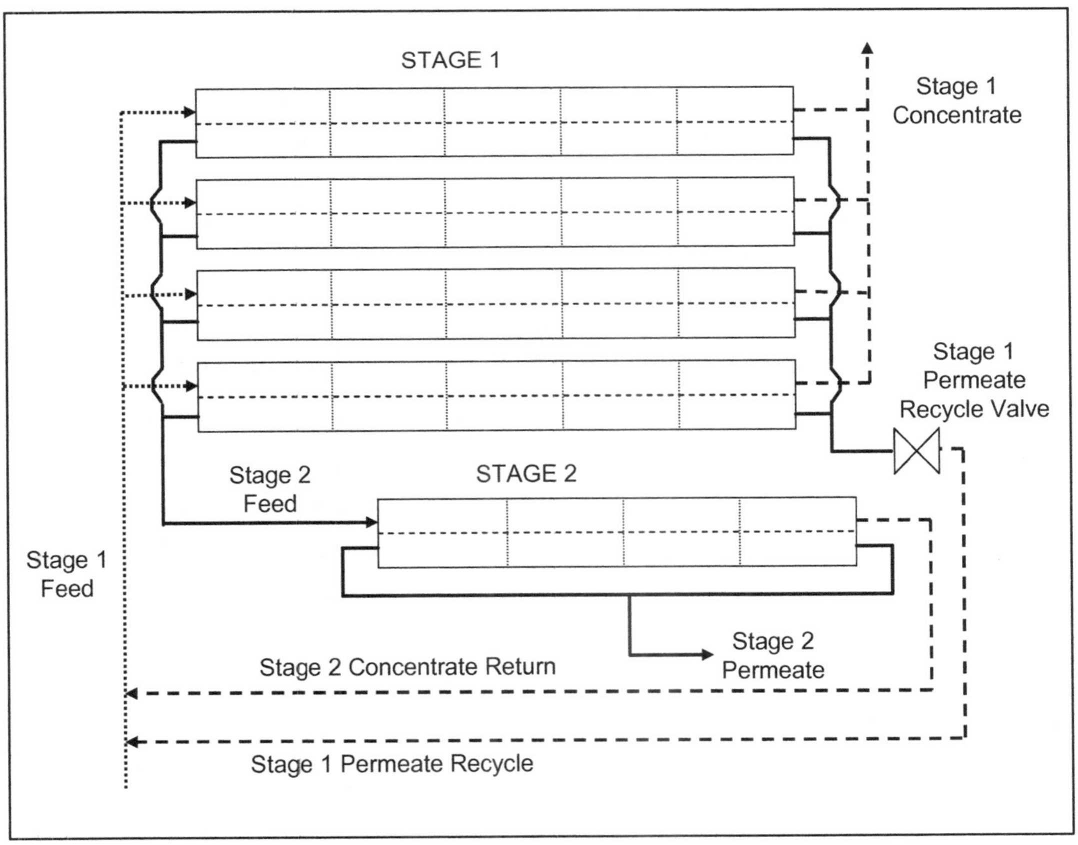

Figure 2.1 Configuration of pilot plant

For each membrane, the optimization test, which lasted 3 weeks, was followed by an approximately one-week test to verify reproducibility, assess process performance at the optimal conditions on a longer duration, and determine boron removal through pH adjustment in the second stage.

PILOT TESTS

Results

Tables 2.1 and 2.2 show excerpted results from the 100 pilot tests conducted for NF-90 and NE-90, each at various temperatures (60 to 80°F), Stage 1 applied pressures, percentages of Stage 1 permeate return (0, 20, and 40 percent of total Stage 1 permeate), and Stage 2 permeate salinity goals (200 and 350 mg/L TDS). Throughout the experimental runs, the permeate salinity goals determined Stage 2 applied pressures. Pilot test data related to Tables 2.1 and 2.2 can be found in Appendix B (Tables B.1 through B.8 and B.15 through B.20) and include the following parameters:

- Temperature
- Pressure
- Flow
- pH
- Alkalinity
- Boron
- Silica
- Cations (calcium, magnesium, sodium, potassium)
- Anions (sulfate, chloride, bromide, fluoride, iodide)
- Conductivity
- Total dissolved solids
- Langelier saturation index

Table 2.1
Selected pilot test results for FilmTec NF-90 at average temperature 67°F (19.4°C)

Stage 1 pressure (psi)	Stage 1 permeate recycled (percent)	Stage 2 pressure (psi)	Stage 1 feed salinity (mg/L)	Stage 2 permeate salinity (mg/L TDS)	Recovery (percent)	Energy (kWh/kgal)
580	0	170	28,071	180	39	16.9
		210	31,440	328	41	14.3
	40	170	29,458	176	33	16.7
		210	29,920	331	34	16.4
520	0	170	31,770	206	29	15.9
		220	34,478	380	31	13.8
	40	145	29,392	194	26	18.4
		190	30,185	349	26	16.6

Table 2.2
Selected pilot test results for Saehan NE-90 at average temperature 70°F (21.1°C)

Stage 1 pressure (psi)	Stage 1 permeate recycled (percent)	Stage 2 pressure (psi)	Stage 1 feed salinity (mg/L)	Stage 2 permeate salinity (mg/L TDS)	Recovery (percent)	Energy (kWh/kgal)
520	0	200	24,702	338	42	19.7
	40	140	25,825	176	39	19.1
		200	28,996	355	40	16.0
470	0	170	25,429	364	33	20.3
	40	125	26,631	184	33	19.8
		160	30,237	363	33	16.7

For both membranes, pressure and array configuration were the main factors in determining flow and salinity of the final permeate. For the same Stage 1 pressure and amount of Stage 1 permeate being recycled, the higher permeate salinity target (i.e., 350 mg/L TDS) in Stage 2 was consistently obtained by increasing the Stage 2 pressure. Overall recovery of the system (i.e., the ratio of permeate flow from Stage 2 to raw water flow) appears to be highly dependent on the pressure in Stage 1. For both membranes, a decrease in Stage 1 pressure accompanies a decrease in overall recovery. As the percentage of Stage 1 permeate flow being returned to the feed increases, overall recovery appears to diminish for NF-90 and to remain unchanged for NE-90. The energy in Tables 2.1 and 2.2 is calculated as:

$$\frac{\text{kWh}}{1{,}000\ \text{gal}} = \left(\frac{0.7457}{1714}\right)\left(\frac{1}{60\,\dfrac{Q_{2P}}{1000}}\right)\left[\frac{Q_{1F}P_{1F} + Q_{1P}P_{1P}}{75\%} - 80\%\ Q_{1C}P_{1C} - Q_{2C}P_{2C}\right] \qquad (2.1)$$

with flow expressed in gpm, and pressure in psi. A pump efficiency of 75 percent and an energy recovery efficiency of 80 percent were assumed. Because Stage 2 concentrate is recombined with the feed, there is no energy exchange (hence no efficiency factor for that term in the equation). For simplification, line losses were omitted from Equation (2.1).

Energy levels are about in the same range for both membranes. Assuming a 30 percent recovery in each stage of a brine-staging SWRO system (i.e., the concentrate of the first pass is treated in a second stage, and both permeates are mixed resulting in an overall 51 percent recovery) and operating pressures of about 600 and 850 psi for the respective stages, and using the same efficiency factor assumptions as in Equation (2.1), the corresponding energy level is about 18 kWh/kGal. This means that some conditions for the NF^2 process are not favorable in terms of energy if the calculated energy is greater than 18 kWh/kGal.

As anticipated, the calculated energy increases as the salinity target in Stage 2 decreases. However, the decrease in energy associated with higher Stage 2 pressure (for a same Stage 1 applied pressure and percentage of Stage 1 permeate recycle) is counterintuitive. As higher pressure is applied in Stage 2, the recovery in Stage 2 increases, which decreases the Stage 2 concentrate flow. In turn, this means there is less feedwater to treat in Stage 1 because less

Stage 2 concentrate flow is returned. Because Stage 1 operates under a substantially higher pressure than Stage 2 (about double that of Stage 2) any decrease in feed flow will therefore result in decrease in the calculated energy. The same effect occurs with the percentage of Stage 1 permeate being recycled. Overall, the anticipated energy saved from treating diluted feedwater does not compensate the energy lost from treating the corresponding higher feed flow.

Both NF membranes rejected a greater proportion of divalent ions compared to monovalent ions, essentially shifting the proportion of sodium and chloride ions from about 87 percent (w/w) in the raw water to greater than 95 percent in the final permeate. Additionally, NF membranes are "looser" than SWRO membranes and therefore feed concentration increases less as it progresses along the membrane element, especially at the membrane surface (i.e., the concentration polarization modulus is less for NF than it is for SWRO). The combination of diminished concentration polarization and preference for divalent ion removal alters the supersaturation conditions of sparingly soluble salts so that antiscalant dosing in the NF^2 process can be substantially trimmed down from dosages typically used in SWRO. It also makes the process particularly well suited for the strategy of increasing pH in the second stage to remove boron because the risk of scaling is diminished as compared to SWRO.

Figure 2.2 shows that boron removal improved as pH in Stage 2 was increased, especially beyond pH 9.8 when boron has fully hydrolyzed to borate. At pH 10, both membranes rejected approximately 80 percent of the raw water boron concentration (4.5 mg/L on average) thus resulting in a permeate boron concentration below California's Notification Level (1 mg/L). At lower pH, NF-90 outperformed NE-90, with boron removal levels of 50 to 60 percent for NF-90 compared to 30 to 40 percent for NE-90. The marked "jump" in boron rejection observed for NE-90 around pH 9.8 may be attributed to NE-90's relative "looseness" compared to NF-90, which increases the contrast between boric acid rejection (pH<9.5) and borate (pH>9.5). Boron removal pilot data are included in Appendix B (Tables B.10 to B.15 and B.21 to B.26).

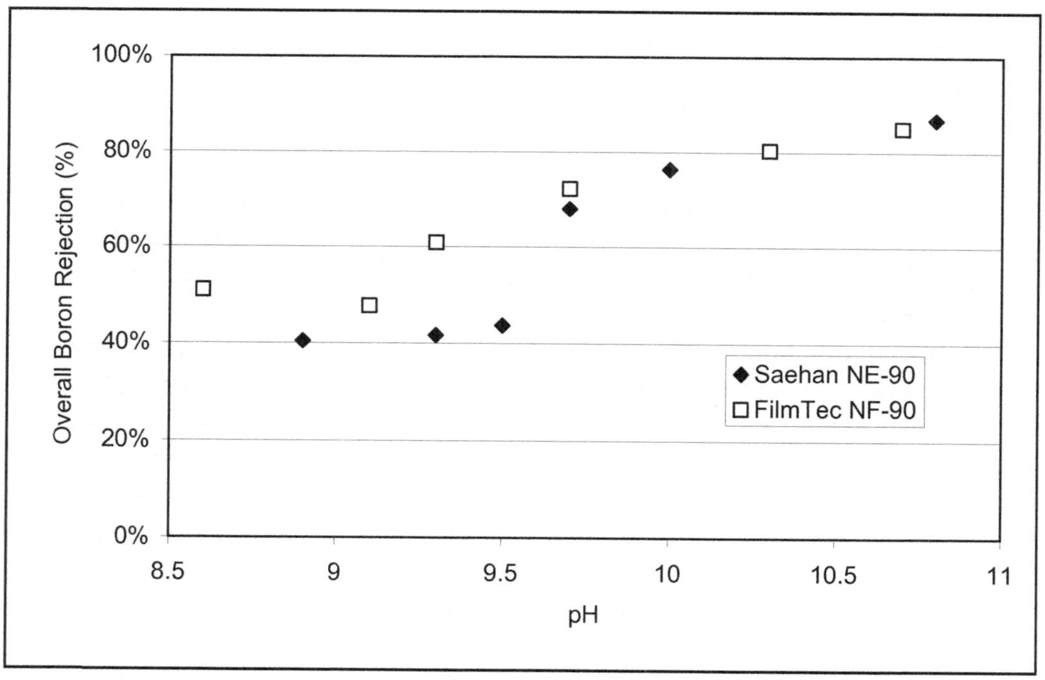

Figure 2.2 Selected pilot test results for boron removal at average temperature 70°F (21.1°C) (boron feed concentration range: 4.6 to 4.8 mg/L)

Based on the pilot data included in Appendix B, Table 2.3 below presents average removal levels achieved by both membranes for fluoride, bromide, silica, and iodide. Removal levels tend to increase with increasing atomic mass unit of the ionic species, with iodide (126.9 g/mol) removed on average at more than 99 percent and fluoride (19 g/mol) with a minimum of 60 percent average removal. This behavior is consistent with tight NF membranes for which removal is partly based on size exclusion.

Tables 2.4 to 2.8 present the permeability and reflection coefficients for NF-90 and NE-90 at the various pilot testing conditions. The k_i and σ values obtained from the pilot data are comparable to the bench-scale results; however the k_w values are substantially lower, although still in the same order of magnitude (see Table 1.5).

For both membranes, the k_w values in Stage 1 were only dependent on that stage operating pressure and slightly decreased with increasing pressure. For Stage 2, the dependence of the water permeability coefficient on test parameters was more varied:

- NE-90 was dependent on the amount of Stage 1 permeate being returned, while NF-90 was unaffected by it
- For both membranes temperature and the final salinity goal played a role with opposite effects: For NF-90 any increase in these factors increased the k_w value in Stage 2, whereas for NE-90 it would decrease it.

The reflection coefficient, σ, for both membranes at either stage was dependent on the amount of Stage 1 permeate being returned, except for NF-90's Stage 1 where it was only dependent on temperature. In general, the higher the portion of Stage 1 permeate being recycled, the lower the reflection coefficient. As for the k_w values, the final salinity target affected σ in Stage 2 (higher salinity lowered σ for both membranes).

Obviously, and as observed in the benchscale tests, both membranes exhibited sharp differences in the k_i values between monovalent and divalent ions. As anticipated (based on the predominance of monovalent ions), the mass transfer (or permeability) coefficient for the overall salinity, k_{TDS}, had a value close to that of the monovalent ions. For NF-90, the k_i value was dependent on temperature and pressure in Stage 1, and on target salinity and percentage of Stage 1 permeate return in Stage 2. For NE-90, the dependence was more complex, with values in both stages being affected by target salinity and the amount of Stage 1 permeate recycle, but additionally values in Stage 1 were also dependent on temperature, whereas in Stage 2 they were also impacted by the operating pressure in Stage 1.

Table 2.3
Average removal for fluoride, bromide, silica, and iodide by NF-90 and NE-90

Membrane	Fluoride (mg/L)		Bromide (mg/L)		Silica (mg/L)		Iodide (μg/L)	
	Raw	Stage 2 permeate	Raw	Stage 2 permeate	Raw	Stage 2 permeate	Raw	Stage 2 permeate
NF-90	0.6	≤0.1	60	≤1	1.2	≤0.1	50	0.37
NE-90	0.6	≤0.1	60	≤1	0.8	≤0.1	60	0.45

Table 2.4
Dependence of water permeability coefficient (k_w) on Stage 1 pressure, target salinity, and temperature for NF-90 and NE-90 based on pilot tests

	Stage 1 k_w (gfd/psi)			Stage 2 k_w (gfd/psi)				
					200 mg/L		350 mg/L	
Membrane	470 psi	520 psi	580 psi	No recycle of Stage 1 Permeate	69°F	73°F	69°F	73°F
NF-90	----	0.029	0.025	----	0.075	0.085	0.092	0.103
NE-90	0.043	0.038	----	0.104	0.123	0.115	0.107	0.099

Table 2.5
Dependence of the TDS-based reflection coefficient (σ) on target salinity, Stage 1 permeate recycle, and temperature for NF-90 and NE-90 based on pilot tests

	Stage 1 σ (percent)					Stage 2 σ (percent)					
						200 mg/L			350 mg/L		
Membrane	0 percent	20 percent	40 percent	66°F	71°F	0 percent	20 percent	40 percent	0 percent	20 percent	40 percent
NF-90	---	---	---	72	75	83	78	74	76	69	63
NE-90	61	57	61	---	---	84	83	79	81	79	73

Table 2.6
Dependence of NF-90 salt permeability coefficient (k_i) on Stage 1 pressure, temperature, target salinity, and Stage 1 permeate recycle based on pilot tests

	Stage 1 k_i (gfd)				Stage 2 k_i (gfd)					
	520 psi		580 psi		200 mg/L			350 mg/L		
					0	20	40	0	20	40
Salt Type	66°F	71°F	66°F	71°F	percent	percent	percent	percent	percent	percent
TDS	0.61	0.78	0.71	0.90	1.05	1.54	1.82	2.04	3.12	3.72
Monovalent	0.88	0.95	1.01	1.22	1.19	1.40	1.97	2.01	3.07	4.24
Divalent	0.10	0.11	0.12	0.12	0.55	0.60	0.54	0.56	0.55	0.56

Table 2.7
Dependence of NE-90 Stage 1 salt permeability coefficient (k_i) on target salinity, Stage 1 permeate recycle, and temperature based on pilot tests

	200 mg/L					350 mg/L					
	0	20		40		0		20		40	
	percent	percent		Percent		percent		percent		percent	
Salt type	64°F	64°F	69°F	64°F	69°F	64°F	69°F	64°F	69°F	64°F	69°F
TDS	2.16	2.28	3.13	1.93	2.28	1.83	2.64	2.29	2.61	1.97	2.24
Monovalent	2.57	2.60	3.80	2.82	3.01	2.05	3.85	2.72	3.22	2.74	3.09
Divalent	0.24	0.24	0.37	0.28	0.28	0.22	0.35	0.27	0.29	0.28	0.29

Table 2.8
Dependence of NE-90 Stage 2 salt permeability coefficients (k_i) on target salinity, Stage 1 pressure, and Stage 1 permeate recycle based on pilot tests

	200 mg/L						350 mg/L					
	470 psi			520 psi			470 psi			520 psi		
Salt type	0 percent	20 percent	40 percent	0 percent	20 percent	40 percent	0 percent	20 percent	40 percent	0 percent	20 percent	40 percent
TDS	0.77	0.88	1.00	0.85	1.10	1.34	1.12	1.46	1.94	1.50	1.90	2.66
Monovalent	0.78	0.88	0.99	0.83	1.01	1.31	1.13	1.78	1.99	1.34	1.97	2.66
Divalent	0.51	0.65	0.78	0.72	0.79	0.96	0.46	0.64	0.77	0.58	0.93	1.05

PREDICTIVE MODEL

Comparison with Pilot Test Results

Tables 2.9 and 2.10 show the model results obtained for FilmTec NF-90 and Saehan NE-90, respectively, based on the same operating parameters in Tables 2.1 and 2.2 for the pilot tests. The excellent fit between the model and the pilot results extends beyond the data presented in Tables 2.9 and 2.10, with less than 10 percent error on determining the final permeate flow and salinity, regardless of operating conditions (e.g., temperature, pressure in Stage 1 or Stage 2, Stage 1 permeate recycle, etc.).

During model calibration, equations relating the reflection and permeability coefficients in each stage to system parameters (i.e., raw water flow, pressure, temperature, and amount of Stage 1 permeate recycle) were developed for both membranes. A sensitivity analysis determined that an overall reflection coefficient (instead of one for each stage) was sufficient to provide a good match between model and pilot test results.

Table 2.9
Selected model results for FilmTec NF-90 at average temperature 67°F (19.4°C)

Stage 1 pressure (psi)	Stage 1 permeate recycled (percent)	Stage 2 pressure (psi)	Final salinity (mg/L TDS)	Stage 2 permeate (gpm)
580	0	170	187	3.9
		210	335	4
	40	170	192	4.1
		210	345	41 4.3
520	0	170	204	3.6
		220	373	4.3
	40	145	192	3.3
		190	361	3.6

Table 2.10
Selected model results for Saehan NE-90 at average temperature 70°F (21.1°C)

Stage 1 pressure (psi)	Stage 1 permeate recycled (percent)	Stage 2 pressure (psi)	Final salinity (mg/L TDS)	Stage permeate (gpm)
520	0	200	384	4.1
	40	140	206	3.8
		200	378	4.9
470	0	170	312	3.5
	40	125	144	3.3
		160	328	3.9

Correlations for the permeability and reflection coefficients of each membrane are listed below. The equations display functions of system parameters consistent with the trends observed in Tables 2.4 through 2.8. Parameters used for the correlations are:

- Q_{raw}: raw water flow (gpm)
- T: temperature (°F)
- P: applied pressure (psi)
- %: percentage of Stage 1 permeate return (e.g., entered as 0.4 for 40 percent)
- σ: overall reflection coefficient
- k_w: water permeability coefficients (gfd/psi)
- k_i: permeability coefficient for salt i (gfd)
- Subscripts 1 and 2 denote Stage 1 and Stage 2, respectively

Correlated Coefficients for NF-90

For NF-90, the equations and conditions for the coefficients are:

$$k_{w,1} = 0.0003\,T + 0.017 \tag{2.2}$$

$$k_{w,2} = 0.002\,T + 0.05 \tag{2.3}$$

$$\sigma = (3.3 + \%)\frac{P_1}{10000} + 0.69 \tag{2.4}$$

$$k_{i,1} = 0.0017\,P_1 + 0.25 \tag{2.5}$$

Depending on whether a portion of Stage 1 permeate is recycled, k_i in Stage 2 is governed by either of the following equations:

$$\% = 0 : k_{i,2} = 0.0009\,(P_2\,(6.8 - 0.08\,Q_{raw} - 0.0033\,T) - 470) + 0.3425 \tag{2.6}$$

$$\% \neq 0 : k_{i,2} = 0.0016\,(P_2\,(0.09\,T - 2.8) + 300\,(\%) - 460) + 0.2 \tag{2.7}$$

Correlated Coefficients for NE-90

NE-90's equations and conditions for the reflection and permeability coefficients are as follows:

$$k_{w,1} = 0.015\,(\%) + 0.055 \tag{2.8}$$

$$k_{i,1} = 2.5 - \% \tag{2.9}$$

Depending on whether recycle of part of the Stage 1 permeate occurs, one of the following two equations should be applied for k_w in Stage 2:

$$\% = 0 : k_{w,2} = (0.0675 - 0.000154\, P_1)\, T + 0.01004\, P_1 - 4.25 \quad (2.10)$$

$$\% \neq 0 : k_{w,2} = 0.437 - 0.00425\, T \quad (2.11)$$

Depending on temperature and whether partial recycle of Stage 1 permeate takes place, k_i in Stage 2 is governed by one of the five following equations:

$$\% = 0 : k_{i,2} = 0.0041\, P_2 + 0.1\, T - 6.583 \quad (2.12)$$

$$\% < 0.3 \text{ and } T < 66^\circ\text{F} : k_{i,2} = 0.0082\, P_2 - 0.05781\, (\%)\, T \quad (2.13)$$

$$\% \geq 0.3 \text{ and } T < 66^\circ\text{F} : k_{i,2} = 0.0082\, P_2 - 0.01325\, (\%)\, T \quad (2.14)$$

$$\% < 0.3 \text{ and } T \geq 66^\circ\text{F} : k_{i,2} = 0.0082\, P_2 - 0.03146\, (\%)\, T \quad (2.15)$$

$$\% \geq 0.3 \text{ and } T \geq 66^\circ\text{F} : k_{i,2} = 0.0082\, P_2 - 0.00061\, (\%)\, T \quad (2.16)$$

System Trends Based on Predictive Model Results

Figures 2.4 through 2.9 present the results from the predictive model for the following performance indicators: energy, recovery, and target salinity. Figures 2.3 and 2.4 show that a low salinity target (e.g., 200 mg/L) requires more energy than a higher salinity goal (e.g., 350 mg/L). This is particularly obvious for NE-90, because being "looser" than NF-90 it processes more flow than NF-90 to reach a low salinity target. The increased flow for NE-90 comes from the low recovery experienced in Stage 2 – being a "looser" membrane, recovery for NF-90 in that stage cannot be high without compromising permeate quality –, which results in relatively high Stage 2 concentrate flows being returned to the Stage 1 feed. Therefore, because of the larger flows being processed, and in spite of comparatively lower operating pressures, the required energy (i.e., flow multiplied by pressure) for NE-90 remains higher than for NF-90.

Conversely, the "tightness" of the NF-90 results in a much narrower spread of the required energy when the final salinity target varies from 350 to 200 mg/L TDS. However, NF-90's higher selectivity also means that the effect of partial recycle of Stage 1 permeate is detrimental in terms of energy (Figure 2.3) and recovery (Figure 2.5), because more flow is processed but without significant benefit from feed dilution. On the other hand, partial recycle of Stage 1 permeate is beneficial in terms of energy (Figure 2.4) and recovery (Figure 2.6) for NE-90, especially at low salinity targets, because it requires a lesser amount of that recycle flow than of the Stage 2 concentrate flow to benefit from the dilution of the Stage 1 feed salinity (i.e., the salinity of Stage 2 concentrate flow is higher than that of the portion of Stage 1 permeate being returned).

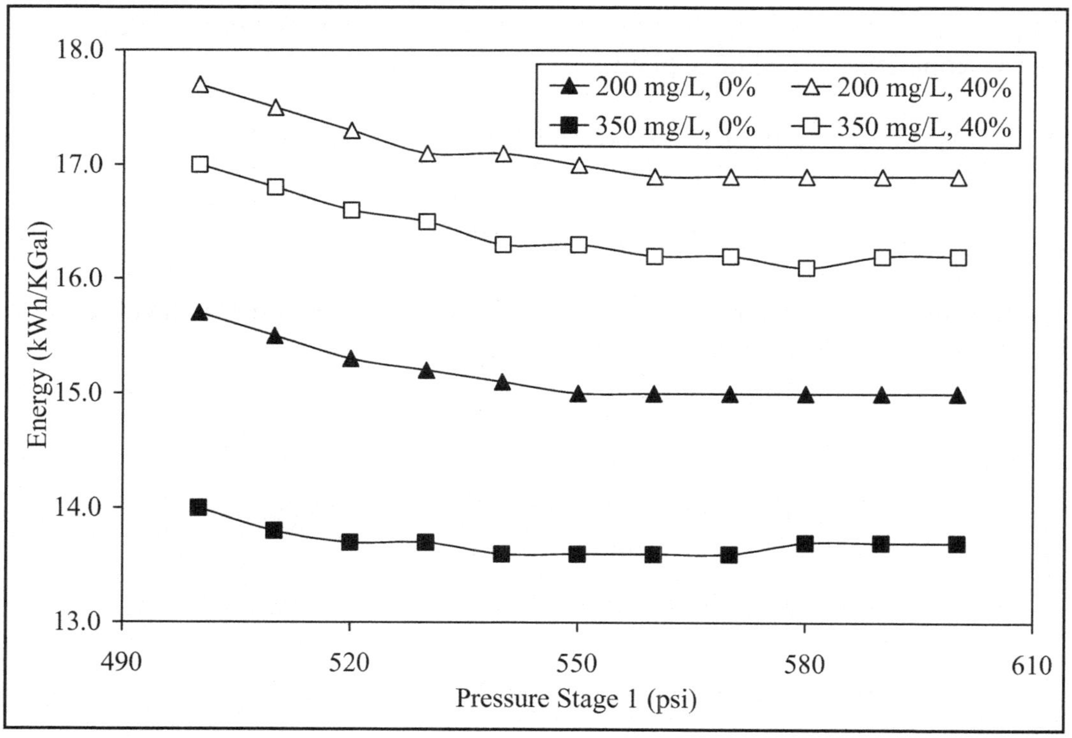

Figure 2.3 NF-90 energy requirements based on Stage 1 pressure for two salinity goals (200 and 350 mg/L TDS) when part of Stage 1 permeate is recycled (0 and 40 percent)

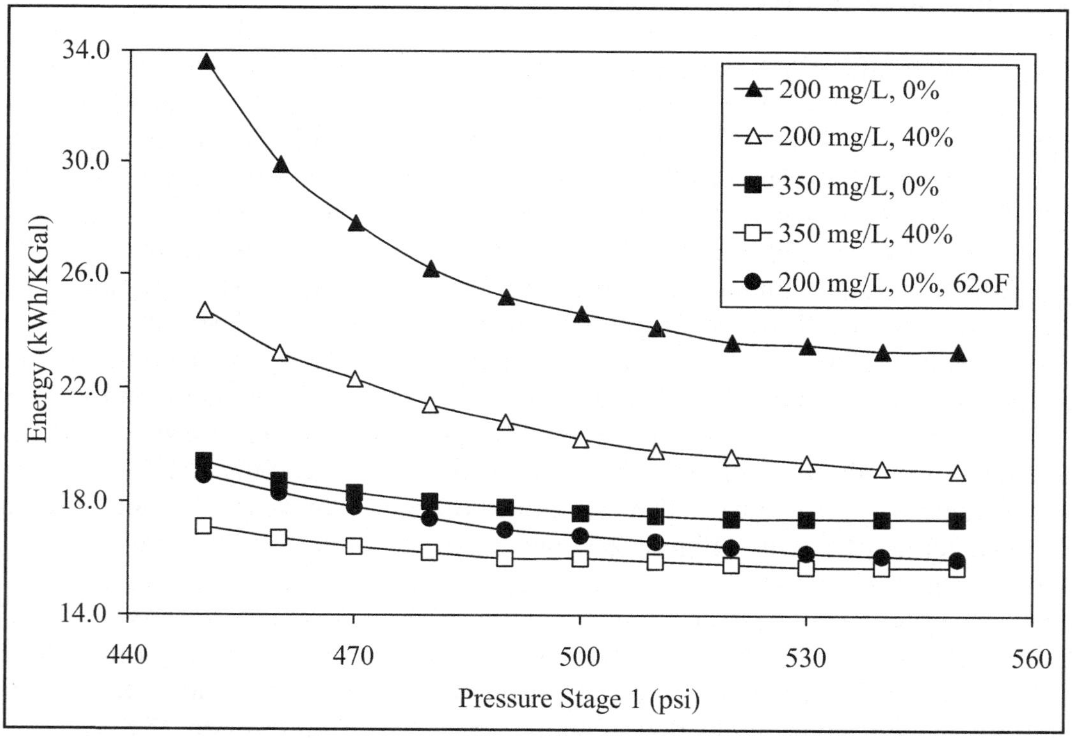

Figure 2.4 NE-90 energy requirements based on Stage 1 pressure for two salinity goals (200 and 350 mg/L TDS) when part of Stage 1 permeate is recycled (0 and 40 percent)

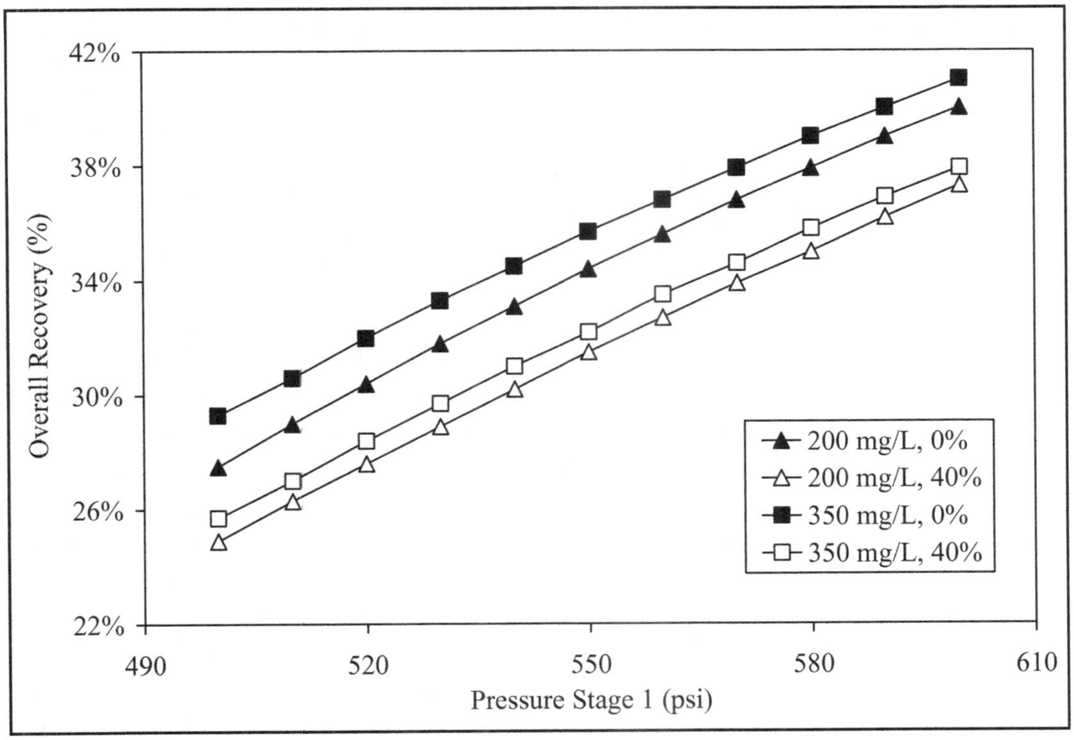

Figure 2.5 NF-90 recovery based on Stage 1 pressure for two salinity goals (200 and 350 mg/L TDS) and when part of Stage 1 permeate is recycled (0 and 40 percent)

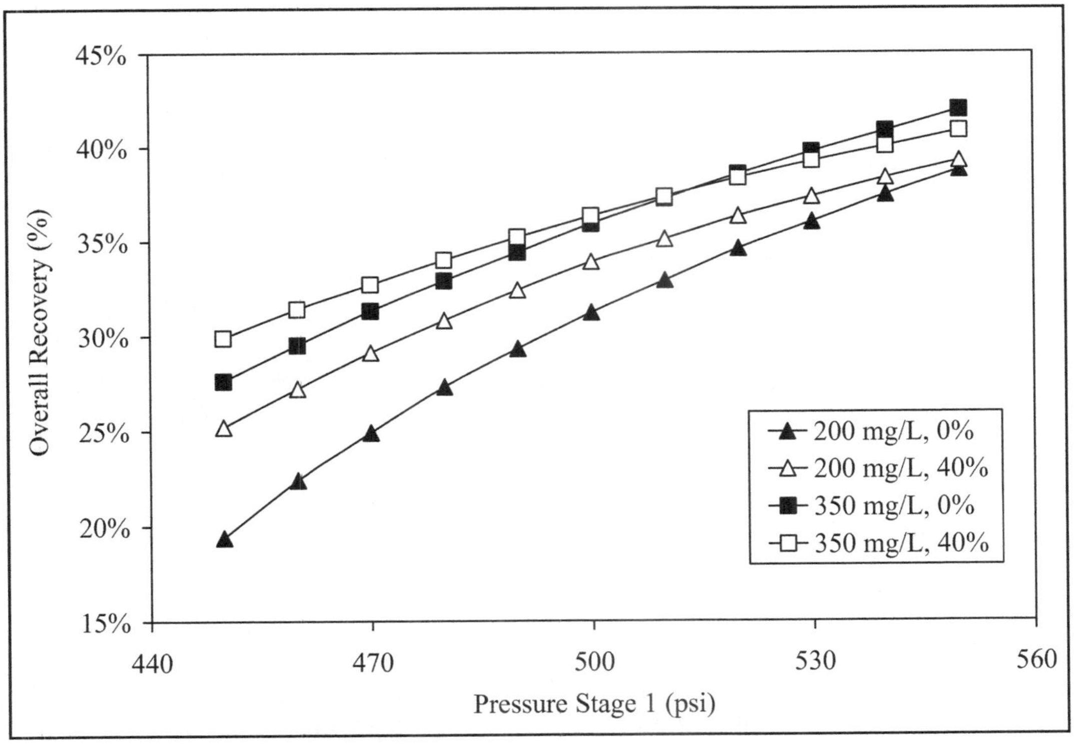

Figure 2.6 NE-90 recovery based on Stage 1 pressure for two salinity goals (200 and 350 mg/L TDS) and when part of Stage 1 permeate is recycled (0 and 40 percent)

For both membranes, pressure in Stage 2 essentially dictates the final salinity. Less pressure is required when Stage 1 permeate is partially recycled because the feed to Stage 2 becomes less saline, and therefore there is less osmotic pressure to overcome. Figure 2.7 also shows that as pressure increases, so does salinity for both membranes. Whereas the relationship between water flux and pressure remains linear, the salt concentration on the membrane surface increases exponentially with pressure, so that beyond a certain transmembrane pressure the salt flux becomes greater than the water flux, resulting in higher salinity in the permeate and therefore a lower salt rejection.

Figures 2.8 and 2.9 neatly illustrate that the bulk of the energy required to operate the system can be attributed to Stage 1. Applied pressure and feed flow in Stage 1 are usually at least twice those of Stage 2, so that the hydraulic power for Stage 1 would be at least 4 times that for Stage2. In other words, Stage 1 contributes at least 80 percent of the total energy required. Figures 2.8 and 2.9 also underline the fact that any effort in saving substantial energy should focus on Stage 1 rather than Stage 2. NF-90 appears to be more suitable than NE-90 for the NF^2 process because the energy requirement is overall significantly less than for NE-90 and remains comparatively unaffected by different salinity targets (i.e., less energy variations based on final permeate salinity goals). Finally, if NF-90 (or an NF membrane with similar characteristics) is used in LBWD's novel seawater desalination process, partial recycle of the Stage 1 permeate should be avoided because it increases the energy requirement (and recovery) without any significant gain in permeate quality.

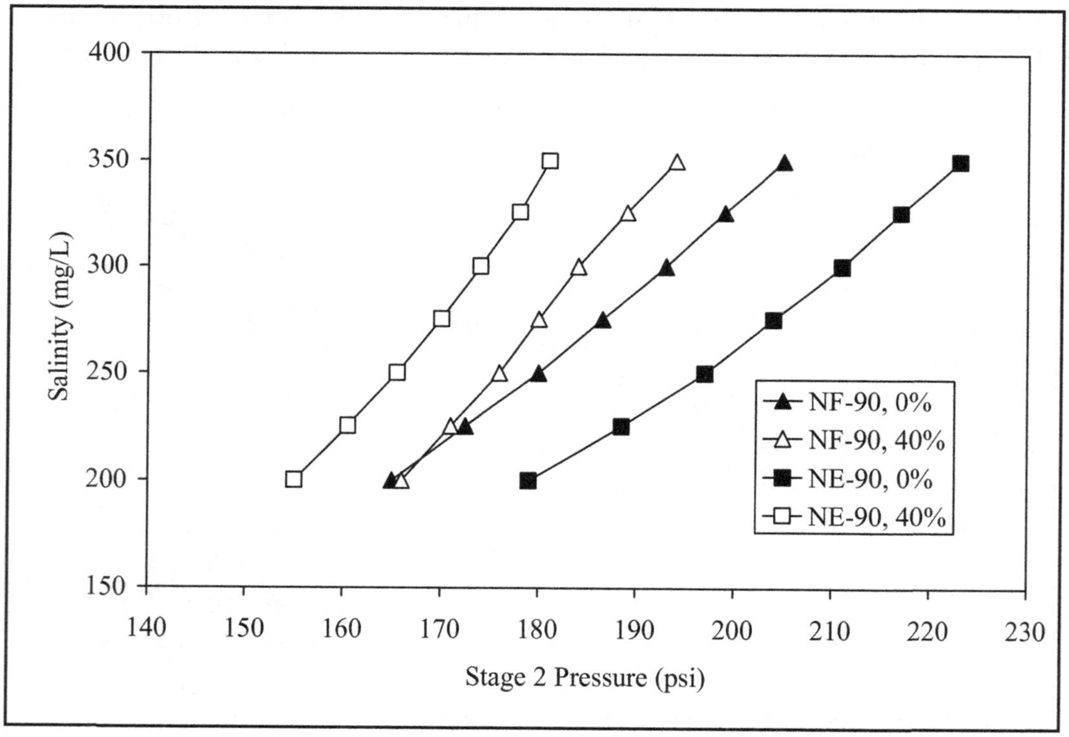

Figure 2.7 NF-90 and NE-90 permeate salinity based on Stage 2 pressure when part of Stage 1 permeate is recycled (0 and 40 percent). Stage 1 pressure: 520 psi

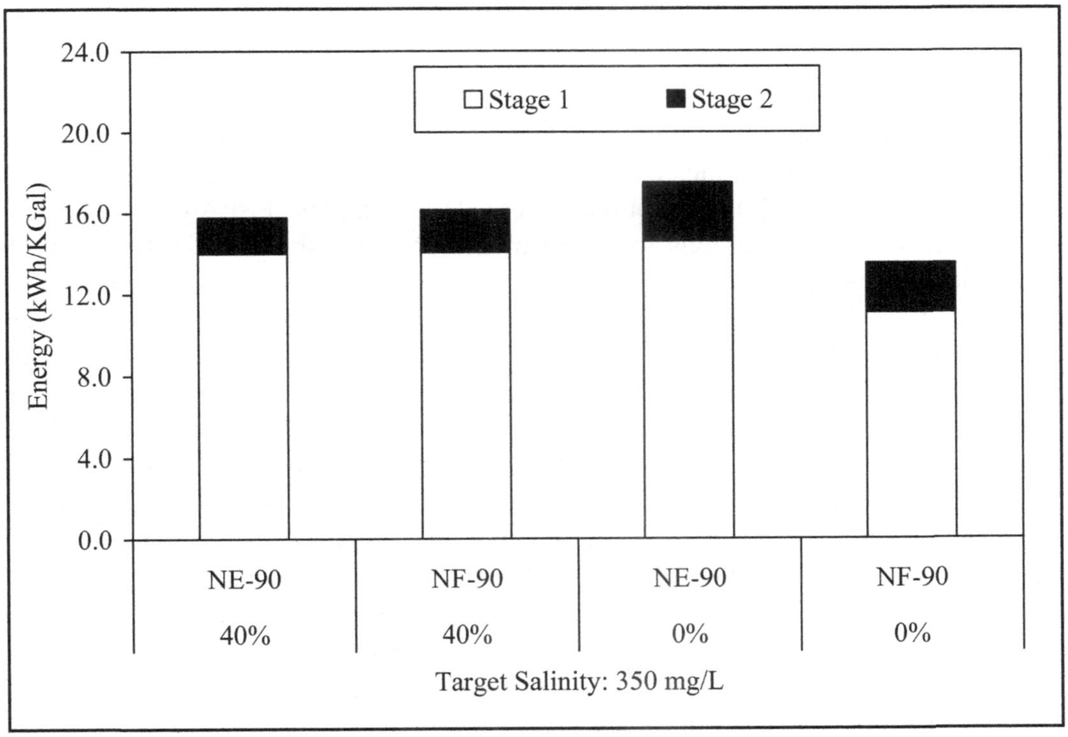

Figure 2.8 NF-90 and NE-90 stage energy contribution to meet salinity target of 350 mg/L TDS when part of Stage 1 permeate is recycled (0 and 40 percent). Recovery: 37 percent

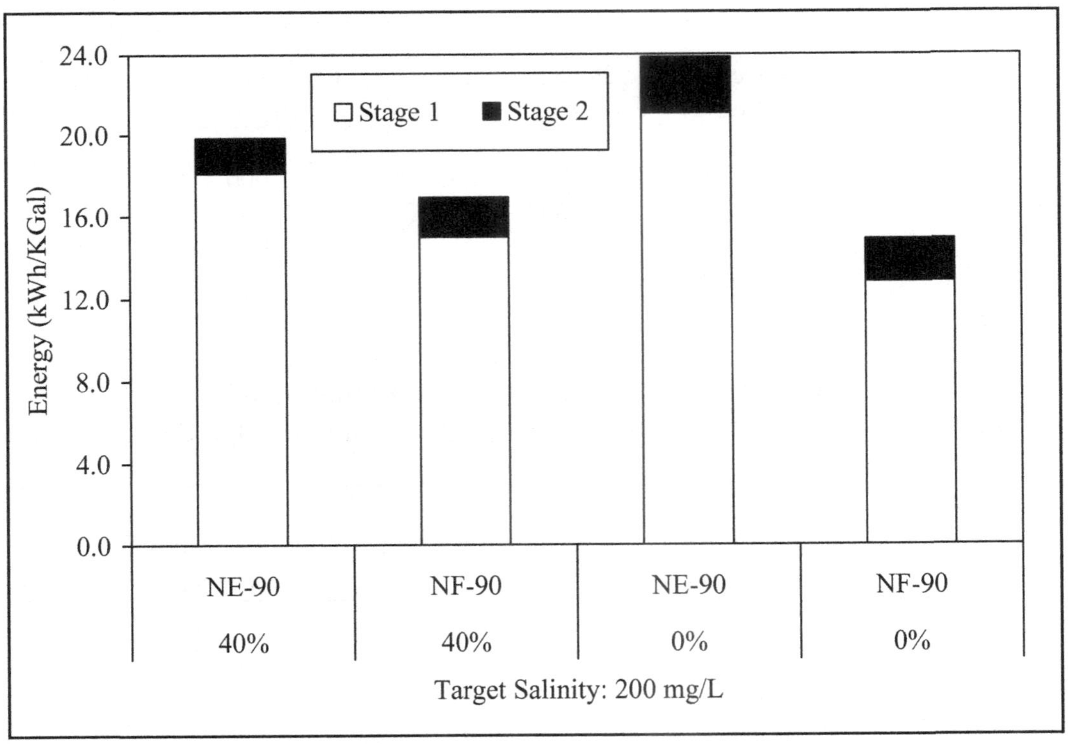

Figure 2.9 NF-90 and NE-90 stage energy contribution to meet salinity target of 200 mg/L TDS when part of Stage 1 permeate is recycled (0 and 40 percent). Recovery: 37 percent

Mixed Membrane Tests

Various combinations of NF and low pressure RO membranes were pilot-tested to evaluate the benefits, if any, of mixing and matching membranes with different selectivity in Stage 1 and Stage 2. NF membranes tested included FilmTec NF-90, TriSep TS80, Saehan NE-90 and the low pressure RO membranes were Hydranautics ESPA4 and TriSep ACM. Manufacturer rejection specifications for these membranes are shown in Table 2.11. Selected results for various combinations are shown in Table 2.12. The increasing "tightness" order of the membranes is: TS-80 < NE-90 < ESPA4 < NF-90 < ACM5. Pilot tests were conducted with the various "looser" membranes in the front end (Stage 1) and "tighter" membranes in the last stage (Stage 2) to determine if any such combination would further improve the energy and/or recovery of the system. The target salinity was the same for all tests at 200 mg/L TDS.

The results from Table 2.12 show that as membranes in Stage 1 get "tighter" (e.g., NF-90 *versus* TS-80) the calculated energy decreases. This is also consistent with a decreasing Stage 1 permeate recycle flow. High pressure in Stage 2 increases recovery in that stage thereby decreasing Stage 2 concentrate flow returned to feed Stage 1. The data from Table 2.12 suggest that optimal combination in terms of low energy (about 18 kWh/kgal) and high overall recovery (about 40 percent) would be NE-90 followed by ACM5 (and 20 percent of Stage 1 permeate recycle). Alternatively, a system with NF-90 membranes in both stages (without Stage 1 permeate recycle) would also be appropriate (see Table 2.1).

Table 2.11
Manufacturer rejection specifications for NF and RO membranes pilot tested

		NaCl			$MgSO_4$		
Manufacturer	Membrane	Concentration (mg/L)	Pressure (psi)	Rejection (percent)	Concentration (mg/L)	Pressure (psi)	Rejection (percent)
Saehan	NE-90	----	----	----	2,000	150	99.0
FilmTec	NF-90	2,000	70	85-95	2,000	70	>97
Hydranautics	ESPA4	500	100	99.2	----	----	----
TriSep	TS-80	----	----	----	2,000	110	99.0
	ACM5	2,000	150	98.5	----	----	----

Table 2.12
Selected pilot test results for combination of
NF and low pressure RO membranes (target permeate salinity: 200 mg/L TDS)

Membrane					
Stage 1	Stage 2	Stage 1 recycle (percent)	Stage 2 pressure (psi)	Overall recovery (percent)	Energy (kWh/kgal)
TS-80	ESPA4	40	190	24	40
TS-80	ACM5	40	270	27	34.4
NE-90	ESPA4	40	140	41	20.5
NE-90	ACM5	20	300	39	18.0
NF-90	ESPA4	0	160	28	16.9
NF-90	ACM5	0	290	27	15.8

Stage 1 pressure: 580 psi, target salinity: 200 mg/L TDS

A membrane not tight enough in the first stage means a permeate of lower quality (e.g., greater than 10,000 mg/L TDS), which results in a tighter membrane required in Stage 2. Although this system would maximize the recovery, it would not be energy efficient. However, a membrane too tight in Stage 1 may not produce enough permeate to feed Stage 2 to reach a satisfactory overall recovery.

OPTIMIZATION MODEL

Background

One aspect that has not been fully addressed is the impact of loading rate on the system performance. Loading rate (Q_f) basically affects the recovery (Q_p/Q_f) of each stage. In turn, the recovery affects the concentration polarization layer on the membrane surface, with usually an increase in recovery resulting in higher salt passage. This means the raw water flow rate affects the system performance, but how substantial is this effect?

Therefore the relationship of recovery (quantity performance) and rejection (quality performance) needs to be modeled to determine the benefit, if any, of mixing membranes with various salt rejection and flux capabilities on the overall system performance. The following section describes the steps for the mathematical development of a simple "mixed membrane model" (M^3), which would be considered at a "macro" level compared to the "micro" level of the "predictive model" (PM) previously developed.

Model Development

The following notations will be useful to follow the mathematical development steps for M^3. Let:

- Q: flow (gpm)
- C: salt concentration (mg/L TDS)
- Y: recovery (ratio of permeate flow to feed flow)
- R: salt rejection
- %: percentage of Stage 1 permeate being returned
- 1, 2: subscripts denoting Stage 1 and Stage 2, respectively
- raw, f, p, b: subscripts denoting raw, feed, permeate, and brine, respectively

The model has three governing principles:

- Recovery in Stage 1 and Stage 2
- Mass balance
- Concentration factor in Stage 1 and Stage 2

Recovery

From the definition of recovery (ratio of permeate flow to feed flow), let:

$$Y_1 = \frac{Q_{p,1}}{Q_{b,2} + \%\, Q_{p,1} + Q_{raw}} \tag{2.17}$$

$$Y_2 = \frac{Q_{p,2}}{Q_{p,1} - \%\, Q_{p,1}} \tag{2.18}$$

From hydraulic balance, equation (2.18) can be expressed as:

$$Y_2 = \frac{Q_{p,1} - \%\, Q_{p,1} - Q_{b,2}}{Q_{p,1} - \%\, Q_{p,1}} = 1 - \frac{Q_{b,2}}{Q_{p,1} - \%\, Q_{p,1}} \tag{2.19}$$

So that:

$$Q_{b,2} = \left(Q_{p,1} - \%\, Q_{p,1}\right)\left(1 - Y_2\right) \tag{2.20}$$

$Q_{b,2}$ can also be expressed using (2.17) as:

$$Q_{b,2} = \frac{Q_{p,1}}{Y_1} - Q_{raw} - \%\, Q_{p,1} \tag{2.21}$$

Combining equations (2.20) and (2.21) yields:

$$\left(Q_{p,1} - \%\, Q_{p,1}\right)\left(1 - Y_2\right) = \frac{Q_{p,1}}{Y_1} - Q_{raw} - \%\, Q_{p,1} \tag{2.22}$$

So that:

$$Q_{raw} = Q_{p,1}\left(\frac{1}{Y_1} - \% - (1 - \%)\left(1 - Y_2\right)\right) = Q_{p,1}\left(\frac{1}{Y_1} - 1 + Y_2\,(1 - \%)\right) \tag{2.23}$$

Equation (2.23) shows that Y_1 and Y_2 depend on the raw water flow rate (Q_{raw})

Mass Balance

From mass balance:

$$C_{f,1} = \frac{C_{b,2}\, Q_{b,2} + C_{\%\,p,1}\, Q_{\%\,p,1} + C_{raw}\, Q_{raw}}{Q_{\%p,1} + Q_{b,2} + Q_{raw}} \tag{2.24}$$

By approximating the salt concentration in the portion of Stage 1 permeate flow being returned as a result of salt concentration in the brine that passed through the membrane (because the portion being recycled corresponds to permeate collected towards the end of the pressure vessel), equation (2.24) can be rewritten as:

$$C_{\mathrm{f},1} = \frac{C_{\mathrm{b},2}\, Q_{\mathrm{p},1}(1-\%)(1-Y_2) + (1-R_1)\, C_{\mathrm{b},1}\,\%\, Q_{\mathrm{p},1} + C_{\mathrm{raw}}\, Q_{\mathrm{raw}}}{\%\, Q_{\mathrm{p},1} + Q_{\mathrm{p},1}(1-\%)(1-Y_2) + Q_{\mathrm{raw}}} \qquad (2.25)$$

Using equation (2.23), equation (2.25) can be rewritten as:

$$C_{\mathrm{f},1} = \frac{C_{\mathrm{b},2}\, Q_{\mathrm{p},1}(1-\%)(1-Y_2) + (1-R_1)\, C_{\mathrm{b},1}\,\%\, Q_{\mathrm{p},1} + C_{\mathrm{raw}}\, Q_{\mathrm{p},1}\left(\frac{1}{Y_1} - 1 + Y_2\,(1-\%)\right)}{\%\, Q_{\mathrm{p},1} + Q_{\mathrm{p},1}(1-\%)(1-Y_2) + Q_{\mathrm{p},1}\left(\frac{1}{Y_1} - 1 + Y_2\,(1-\%)\right)} \qquad (2.26)$$

So that:

$$C_{\mathrm{f},1} = \frac{C_{\mathrm{b},2}\,(1-\%)(1-Y_2) + (1-R_1)\, C_{\mathrm{b},1}\,\% + C_{\mathrm{raw}}\left(\frac{1}{Y_1} - 1 + Y_2\,(1-\%)\right)}{\% + 1 - \% - Y_2(1-\%) + \frac{1}{Y_1} - 1 + Y_2\,(1-\%)} \qquad (2.27)$$

Therefore:

$$\frac{C_{\mathrm{f},1}}{Y_1} = C_{\mathrm{b},2}\,(1-\%)(1-Y_2) + (1-R_1)\, C_{\mathrm{b},1}\,\% + C_{\mathrm{raw}}\left(\frac{1}{Y_1} - 1 + Y_2\,(1-\%)\right) \qquad (2.28)$$

Re-ordering equation (2.28) yields:

$$C_{\mathrm{b},2} = \frac{\frac{C_{\mathrm{f},1}}{Y_1} - \%\,(1-R_1)\, C_{\mathrm{b},1} - C_{\mathrm{raw}}\left(\frac{1}{Y_1} - 1 + Y_2\,(1-\%)\right)}{(1-\%)(1-Y_2)} \qquad (2.29)$$

Concentration Factor

Using Equation (1.10), the concentration factor in Stage 1 and in Stage 2 can be expressed as:

$$\frac{C_{\mathrm{b},1}}{C_{\mathrm{f},1}} = 1 + \frac{Y_1\, R_1}{1 - Y_1} \qquad \text{(Stage 1)} \qquad (2.30)$$

$$\frac{C_{b,2}}{C_{f,2}} = 1 + \frac{Y_2\ R_2}{1 - Y_2} \qquad \text{(Stage 2)} \tag{2.31}$$

Equation (2.30) can be rearranged to yield:

$$1 - \frac{C_{b,1}}{C_{f,1}} = Y_1\left(1 - R_1 - \frac{C_{b,1}}{C_{f,1}}\right) \tag{2.32}$$

So that:

$$Y_1 = \frac{C_{b,1} - C_{f,1}}{C_{b,1} - C_{f,1}(1 - R_1)} \tag{2.33}$$

The salinity of the feed flow in Stage 2 can be approximated as the salt concentration in the Stage 1 permeate on the entry (feed) side of the pressure vessel:

$$\frac{C_{b,2}}{C_{f,2}} = \frac{C_{b,2}}{C_{f,1}(1 - R_1)} \tag{2.34}$$

Plugging equations (2.29) and (2.34) in equation (2.31) yields:

$$\frac{\dfrac{C_{f,1}}{Y_1} - \%\,(1 - R_1)\,C_{b,1} - C_{raw}\left(\dfrac{1}{Y_1} - 1 + Y_2\,(1 - \%)\right)}{(1 - \%)(1 - Y_2)(1 - R_1)\,C_{f,1}} = 1 + \frac{Y_2\ R_2}{1 - Y_2} \tag{2.35}$$

Equaiton (2.35) can be rearranged as:

$$\frac{C_{f,1}}{Y_1} - \%\,(1 - R_1)\,C_{b,1} - C_{raw}\left(\frac{1}{Y_1} - 1 + Y_2\,(1 - \%)\right) = (1 - \%)(1 - R_1)\,C_{f,1}(1 - Y_2(1 - R_2)) \tag{2.36}$$

$$Y_2 = \frac{C_{f,1}\left(\dfrac{1}{Y_1} - (1 - R_1)(1 - \%)\right) - C_{b,1}\,\%\,(1 - R_1) + C_{raw}\left(1 - \dfrac{1}{Y_1}\right)}{(1 - \%)\left(C_{raw} - C_{f,1}(1 - R_1)(1 - R_2)\right)} \tag{2.37}$$

Plugging equation (2.33) into equation (2.37) yields

$$Y_2 = \frac{C_{f,1}\left(\dfrac{C_{b,1} - C_{f,1}(1 - R_1)}{C_{b,1} - C_{f,1}} - (1 - R_1)(1 - \%)\right) - C_{b,1}\,\%\,(1 - R_1) + C_{raw}\left(1 - \dfrac{C_{b,1} - C_{f,1}(1 - R_1)}{C_{b,1} - C_{f,1}}\right)}{(1 - \%)\left(C_{raw} - C_{f,1}(1 - R_1)(1 - R_2)\right)} \tag{2.38}$$

Which can be rearranged as:

$$Y_2 = \frac{C_{\text{f},1}\left(\frac{R_1\, C_{\text{b},1}}{C_{\text{b},1} - C_{\text{f},1}} + \%\,(1 - R_1)\right) - C_{\text{b},1}\%\,(1 - R_1) - C_{\text{raw}}\left(\frac{R_1\, C_{\text{f},1}}{C_{\text{b},1} - C_{\text{f},1}}\right)}{(1 - \%)\left(C_{\text{raw}} - C_{\text{f},1}(1 - R_1)(1 - R_2)\right)} \tag{2.39}$$

Equation (2.39) can be rewritten as:

$$Y_2 = \frac{\left(\frac{R_1\, C_{\text{f},1}}{C_{\text{b},1} - C_{\text{f},1}}\right)(C_{\text{b},1} - C_{\text{raw}}) - \%\,(1 - R_1)(C_{\text{b},1} - C_{\text{f},1})}{(1 - \%)\left(C_{\text{raw}} - C_{\text{f},1}(1 - R_1)(1 - R_2)\right)} \tag{2.40}$$

Y_{overall} is the overall system recovery and is defined as:

$$Y_{\text{overall}} = \frac{Q_{\text{p},2}}{Q_{\text{raw}}} \tag{2.41}$$

Using equations (2.18) and (2.23), equation (2.41) can be rewritten as:

$$Y_{\text{overall}} = \frac{Q_{\text{p},1}\, Y_2\,(1 - \%)}{Q_{\text{p},1}\left(\frac{1}{Y_1} - 1 + Y_2\,(1 - \%)\right)} = \frac{Y_1\, Y_2\,(1 - \%)}{1 - Y_1 + Y_1\, Y_2\,(1 - \%)} = \frac{1}{1 + \frac{1 - Y_1}{Y_1\, Y_2\,(1 - \%)}} \tag{2.42}$$

Plugging equations (2.33) and (2.40) into equation (2.42) yields:

$$Y_{\text{overall}} = \frac{1}{1 + \frac{1 - \frac{C_{\text{b},1} - C_{\text{f},1}}{C_{\text{b},1} - C_{\text{f},1}(1 - R_1)_1}}{\left(\frac{C_{\text{b},1} - C_{\text{f},1}}{C_{\text{b},1} - C_{\text{f},1}(1 - R_1)}\right)\frac{\left(\frac{R_1\, C_{\text{f},1}}{C_{\text{b},1} - C_{\text{f},1}}\right)(C_{\text{b},1} - C_{\text{raw}}) - \%\,(1 - R_1)(C_{\text{b},1} - C_{\text{f},1})}{(1 - \%)\left(C_{\text{raw}} - C_{\text{f},1}(1 - R_1)(1 - R_2)\right)}}} \tag{2.43}$$

Finally,

$$Y_{\text{overall}} = \frac{1}{1 + \frac{R_1\, C_{\text{f},1}}{\frac{R_1\, C_{\text{f},1}\,(C_{\text{b},1} - C_{\text{raw}}) - \%\,(1 - R_1)(C_{\text{b},1} - C_{\text{f},1})^2}{(1 - \%)\left(C_{\text{raw}} - C_{\text{f},1}(1 - R_1)(1 - R_2)\right)}}} \tag{2.44}$$

Findings

With equations (2.33), (2.40) and (2.44) all recoveries can be determined based on:

- Rejection characteristics of membranes in Stage 1 (R_1) and Stage 2 (R_2)
- Raw water flow and salinity
- Percentage of Stage 1 permeate flow being returned
- Salinity for Stage 1 feed and concentrate flows

Given raw seawater flow and salinity of 14 gpm and 35,000 mg/L TDS, respectively, and a final permeate salinity goal of 350 mg/L TDS, Tables 2.13 through 2.16 show how parameters vary for various targeted Stage 1 conditions (rejection level, feed salinity, concentrate salinity, and permeate recycle). As noted in the previous section, Stage 1 feedwater dilution from recycle flows (Stage 2 concentrate and any percentage of Stage 1 permeate) results in higher transmembrane pressure because the osmotic pressure is lower while the applied pressure remains constant. However, the increased water flux in Stages 1 and 2 from the higher transmembrane pressure (the less saline part of Stage 1 permeate is fed in Stage 2 resulting in lower feed osmotic pressure there as well) does not result in increased system productivity because the generated extra flows are used for recycle. Furthermore, any energy recovery from the brine of Stage 1 and 2 does not compensate the increased energy demand from processing higher feed flows at the same applied pressure.

Table 2.13
Recovery and flow results from the mixed membrane model for different rejection levels in Stages 1 and 2. Stage 1 conditions: no permeate recycle, 60,000 mg/L concentrate salinity, and 30,000 mg/L feed salinity

Rejection Stage 1 (percent)	Rejection Stage 2 (percent)	Recovery Stage 1 (percent)	Recovery Stage 2 (percent)	Overall recovery (percent)	Permeate Stage 1 (gpm)	Concentrate Stage 2 (gpm)	Feed Stage 1 (gpm)
82	95	55	59	42	9.9	4.1	18.1
86	93	54	62	42	9.5	3.6	17.6
90	90	53	65	42	9.0	3.2	17.2
94	84	52	68	42	8.7	2.8	16.8

Table 2.14
Recovery and flow results from the mixed membrane model for different rejection levels in Stages 1 and 2. Stage 1 conditions: no permeate recycle, 50,000 mg/L concentrate salinity, and 30,000 mg/L feed salinity

Rejection Stage 1 (percent)	Rejection Stage 2 (percent)	Recovery Stage 1 (percent)	Recovery Stage 2 (percent)	Overall recovery (percent)	Permeate Stage 1 (gpm)	Concentrate Stage 2 (gpm)	Feed Stage 1 (gpm)
82	95	45	53	30	8.0	3.7	17.7
86	93	44	56	30	7.6	3.3	17.3
90	90	43	58	30	7.2	3.0	17.0
94	84	41	61	30	6.9	2.7	16.7

Table 2.15

Recovery and flow results from the mixed membrane model for different rejection levels in Stages 1 and 2. Stage 1 conditions: no permeate recycle, 60,000 mg/L concentrate salinity, and 33,000 mg/L feed salinity

Rejection Stage 1 (percent)	Rejection Stage 2 (percent)	Recovery Stage 1 (percent)	Recovery Stage 2 (percent)	Overall recovery (percent)	Permeate Stage 1 (gpm)	Concentrate Stage 2 (gpm)	Feed Stage 1 (gpm)
82	95	50	72	42	8.1	2.2	16.2
86	93	49	76	42	7.7	1.9	15.9
90	90	48	79	42	7.4	1.5	15.5
94	84	47	83	42	7.1	1.2	15.2

Table 2.16

Recovery and flow results from the mixed membrane model for different rejection levels in Stages 1 and 2. Stage 1 conditions: 20 percent permeate recycle, 60,000 mg/L concentrate salinity, and 30,000 mg/L feed salinity

Rejection Stage 1 (percent)	Rejection Stage 2 (percent)	Recovery Stage 1 (percent)	Recovery Stage 2 (percent)	Overall recovery (percent)	Permeate Stage 1 (gpm)	Concentrate Stage 2 (gpm)	Feed Stage 1 (gpm)
82	95	55	74	40	10.2	2.0	18.6
86	93	54	79	41	9.7	1.5	18.0
90	90	53	84	41	9.2	1.1	17.4
94	84	52	89	41	8.7	0.8	16.9

Based on these findings, conditions that result in low Stage 1 feed flow rate will also result in low energy demand. Stage 1 feed flow rate is minimized when:

- Stage 1 concentrate salinity is maximized (i.e., recovery in Stage 1 is maximized)
- Stage 1 feed salinity is minimally diluted (i.e., recycle to Stage 1 feed is minimized)
- Partial recycle of Stage 1 permeate is avoided
- Tighter membranes are placed in Stage 1 (94 percent rejection) and looser membranes in Stage 2 (84 percent rejection)

The first three bullets indicate a system for which applied pressure in Stage 1 is high, the entire Stage 1 permeate feeds into Stage 2, and Stage 2 is under relatively high pressure (which reduces Stage 2 concentrate flow). Specific pilot tests set up to simultaneously meet these requirements confirmed the validity of these conditions, as shown by the energy values displayed in Table 2.1 (rows 2 and 6: 14.3 kWh/kgal at 580 psi and 13.8 kWh/kgal at 520 psi). The placement and rejection characteristics of the "mixed" membranes that were pilot tested (Tables 2.11 and 2.12) did not correspond to the recommendation of the last bullet, therefore this last condition (i.e., "tighter" membranes in Stage 1 and "looser" membranes in Stage 2) has not been verified in the field.

CHAPTER 3
BLENDING STRATEGIES

BACKGROUND

LBWD's existing finished water varies in composition during the year, with 100 percent of treated groundwater (GW) in summer, and 100 percent of purchased surface water (SW) in winter, the latter being provided either as State Project water (SPW), Colorado River water (CRW), or a blend of the two waters. Based on these conditions, the aim is to determine the percentage of desalinated water (DW) that can be blended while considering the issues of disinfection by-product (DBP) formation, disinfectant residual, and corrosivity.

Table 3.1 presents water quality characteristics for the various waters to be blended. The characteristics for DW vary depending on the permeate salinity and pH targets. One of the two pHs considered represent a condition similar to the water in the distribution system (8.0), and the other represents the condition required if additional boron rejection is necessary (with an interstage pH adjustment to pH 9.8).

BROMIDE

DBP presence in the blends was evaluated on a mass balance basis only, because the formation of total trihalomethanes (TTHM) or five haloacetic acids (HAA5) would be minimal upon chlorination of the permeate since total organic carbon (TOC) in the NF^2 permeate would be very low. However, the presence of bromide ions in the permeate may create a stability problem for the residual disinfectant. Bromide reacts very quickly with free chlorine (which will be used as a primary disinfectant on the permeate for C·T requirements) to form hypobromous acid (HOBr), which means bromide represents a source of chlorine demand (AWWA 1999).

Table 3.1
Water quality characteristics of waters used for the blending evaluation

Water	TDS (mg/L)	pH	Alk (mg/L)	Ca (mg/L)	SO_4 (mg/L)	Cl (mg/L)	Br (mg/L)	LSI	RSI	CCPP
GW	240	7.8	130	25	25	35	0.05	-0.4	8.6	-4.6
SPW	330	8.2	100	35	90	55	0.16	0.0	8.3	-0.4
CRW	650	8.0	125	75	265	90	0.08	0.2	7.7	2.7
DW	200	8.0	3.8	0.9	1.1	120	0.5	-3.1	14.2	-11.0
		9.8	3.8	0.9	1.1	120	0.5	-1.3	12.4	-10.9
	350	8.0	3.7	1.5	1.8	226	1.1	-3.1	14.3	-12.0
		9.8	3.7	1.5	1.8	226	1.1	-1.3	12.5	-11.9

LSI: Langelier Saturation Index – Target range: (corrosive) $-0.5 \leq LSI \leq 0.5$ (scaling)
RSI: Ryznar Saturation Index– Target range: (scaling) $7 \leq RSI \leq 9$ (corrosive)
CCPP: calcium carbonate precipitation potential – Target range: (corrosive) $-5 \leq CCPP \leq 3$ (scaling)

The relative reactivity between bromide and HOBr is similar to the relationship between chloride and HOCl. Like chlorine, bromine (in its hypobromous form) reacts with ammonia to form bromamines (White 1986). Therefore, chloramines and bromamines will be formed when ammonia is added to form the residual disinfectant. A breakpoint phenomenon similar to chlorine occurs when the bromine to ammonia nitrogen molar ratio is 1.5. Tribromamine (NBr_3) is the major species of combined residual bromine present beyond the breakpoint and in the pH range of 7 to 8 it decomposes to HOBr, bromide ions, and nitrogen (N_2). Dibromamine ($NHBr_2$) is very active as compared to mono- and dichloramine, and usually decomposes rapidly, reverting to bromide ion, which terminates the bromine residual (White 1986). As the desalinated water is blended with surface waters containing TOC, dibromamine (and any HOBr resulting from breakpoint bromination) will react with the TOC to form brominated disinfection byproducts.

Because chloramine does not oxidize bromide ions to bromine (White 1986), the depletion of residual disinfectant will stop when $NHBr_2$ and HOBr have decomposed to bromide ions. However, monochloramine has been shown to decompose in presence of bromide ions, and thus disinfectant residual depletion may continue beyond the decomposition of $NHBr_2$ and HOBr, albeit at a lower rate (Trofe et al. 1980). Special modified techniques are available to differentiate brominated residuals from chlorinated residuals, but most commonly used measures of disinfectant residual will not (White 1986).

Because it is likely that HOBr and $NHBr_2$ will react with TOC present in the surface waters used for blending, a possible option would be to "quench" the oxidation demand from the unreacted TOC of the surface waters with a chlorine boost (which would create additional TTHM and HAA5) to prevent brominated residuals from forming brominated DBPs, and avoid disinfectant residual depletion. The detrimental formation of additional DBPs would be balanced by the dilution factor associated with the blending.

CORROSIVITY

Several factors influence the rate of corrosion including pH, temperature, and the presence of dissolved solids and gases. Decreasing the pH of a system increases corrosion because more hydrogen ions become available. When the water temperature is increased, the rate of corrosion is generally increased. Therefore, corrosion rates in the summer are greater than those in the winter. However, colder temperatures allow for more dissolved oxygen, which can also increase corrosion (Buecker 1997 and Lane 1993).

Many indices exist to predict scaling and corrosion tendencies of a system. The ones referenced in this report are the Langelier Saturation Index (LSI), Ryznar Saturation Index (RSI), and Calcium Carbonate Precipitation Potential (CCPP) and predict scaling potential based on the solubility of calcium carbonate. The likelihood of the water to dissolve or precipitate $CaCO_3$ is determined by comparing the actual pH to the pH at which the water should be theoretically saturated with $CaCO_3$.

Corrections can be made for the presence of ion pairs, types of scale other than $CaCO_3$, and scale inhibitors. However, scaling tendency can only give the possibility of a system to corrode, but does not directly correlate to corrosion, because the two are not inversely related. For example, if water has a tendency to dissolve $CaCO_3$ scale, it could possibly corrode, and likewise if it has a tendency to precipitate $CaCO_3$, this could form a protective barrier that would inhibit corrosion (Marangou and Savvides 2001).

The LSI indicates the degree of calcium saturation (Corrosion Doctors 1999). This index calculates the driving force for scale formation in waters with a pH between 6.5 and 9.5, but does not quantitatively provides the extent of $CaCO_3$ precipitation (Plottu-Pecheux et al. 2001). The LSI is calculated using

$$LSI = pH - pH_s \tag{3.1}$$

where pH is the measured pH of the water and pH_s is the pH at saturation in calcite or $CaCO_3$. This saturation pH is dependent upon the salinity and alkalinity of the water and can be determined from

$$pH_s = (9.3 + A + B) - (C + D) \tag{3.2}$$

where,

$$A = \frac{log[TDS] - 1}{10} \tag{3.3}$$

$$B = -13.12 * \log(T + 273) + 34.55 \tag{3.4}$$

$$C = log[Ca^{2+} asCaCO_3] - 0.4 \tag{3.5}$$

$$D = log[alkalinity\_asCaCO_3] \tag{3.6}$$

where TDS is the concentration of total dissolved solids, T is the temperature (°C), Ca^{2+} is the concentration of calcium (mg/L as $CaCO_3$), and alkalinity is the concentration of alkalinity mg/L as $CaCO_3$).

Water with an LSI less than zero tends to dissolve scale. An LSI greater than zero indicates that the water will tend to form scale. This calculation does not account for ionic interactions in the water, and is best for waters with a TDS less than 800 mg/L (Buecker 1997). Also, the LSI does not take into account scale inhibitors, dissolved oxygen, water velocity, or other types of scale such as calcium sulfate or magnesium hydroxide.

The RSI uses similar calculations to the LSI ((Marangou and Savvides 2001) and can be calculated using

$$RSI = (2 * pH_s) - pH \tag{3.7}$$

If the RSI is less than 6.0, the water has a tendency to scale. If the RSI is greater than 6.0, the water has a tendency to dissolve $CaCO_3$. If the RSI is greater than 8.0, then mild steel corrosion becomes an increasing problem. This index is of empirical origin and can be used to estimate scale even in the presence of scale inhibitors such as polyphosphates, phosphonates, and acrylates. Unlike the LSI, which only gives the tendency of a system to form scale, the RSI can be used to determine the amount of scale that will be produced. Within RSI values of 2 to 7, the amount of scale that will be produced is:

$$mg\_scale = \frac{4470}{(2pH_s - pH)^{4.73}} \tag{3.8}$$

Within RSI values of 2 to 4 and in the presence of polyphosphate, the amount of scale that will be produced is:

$$mg\_scale = \frac{631000}{(2pH_s - pH)^{8.65}} \tag{3.9}$$

CCPP gives a measure of the amount of $CaCO_3$ that will precipitate or dissolve in a water system if the water analysis is known. It provides an accurate estimation of the $CaCO_3$ precipitation, but it is a complex calculation that requires the use of a computer program (Lane 1993). The Rothberg Tamburini Windsor (RTW) model (see next section) includes the CCPP calculation. A positive CCPP value means scaling (protective); a water is deemed passive for a CCPP value between 0 and -5; a water is considered mildly corrosive if the CCPP value is between -5 and -10; for a CCPP value of less than -10, the water is corrosive (aggressive).

A combination of the indices listed above should be used in order to gain the clearest picture of how water will react with piping materials. However, empirical indices must not be extrapolated to waters that they were not intended for.

RTW Model

The blending model was developed using a spreadsheet software available from AWWA to estimate corrosion control characteristics of blended waters (Rothberg, Tamburini & Winsor Blending Application Package 4.0, Item number 53042 in AWWA Bookstore), which was enhanced with Microsoft Excel® macros to allow for the analyses of multiple blends. The model can be used either as a guide to determine the optimal blend of waters for minimizing the corrosion potential of the final mix, or to evaluate the possible corrosive effects through increasing the percentage of desalinated water in the final mix. Minimizing the corrosivity of the final blend may not result in minimizing DBP formation, and therefore any model result based on minimizing corrosivity requires an assessment of the DBP levels in the final mix.

Figure 3.1 illustrates the steps to determine the LSI of a specific blend. With four conditions on DW (i.e., 200 and 350 mg/L TDS; and pH 8 and 9.8) for various finished water (FW) mixes composed of specific quantities of GW, SPW, and CRW, all blend ranges were considered. Representative results for final blends are shown in Figures 3.2 through 3.8. As anticipated, the corrosivity differences between pH 8 and pH 9.8 are magnified when the proportion of DW increases relative to FW (i.e., from 25 to 75 percent). As groundwater proportion increases relative to surface water, the resulting blend becomes more corrosive.

The LSI appears to be relatively insensitive to the makeup of surface waters. Only CCPP, and RSI to a lesser degree, seems to capture variations in surface water (SW) make up, with 100 percent SPW being more corrosive than 100 percent CRW. The CCPP is affected by the permeate salinity, with 350 mg/L TDS being more corrosive than 200 mg/L TDS for the same pH values. Of the indices considered, it appears that CCPP shows the greatest sensitivity to the various water quality variations, and is therefore more useful than the other indices in this study.

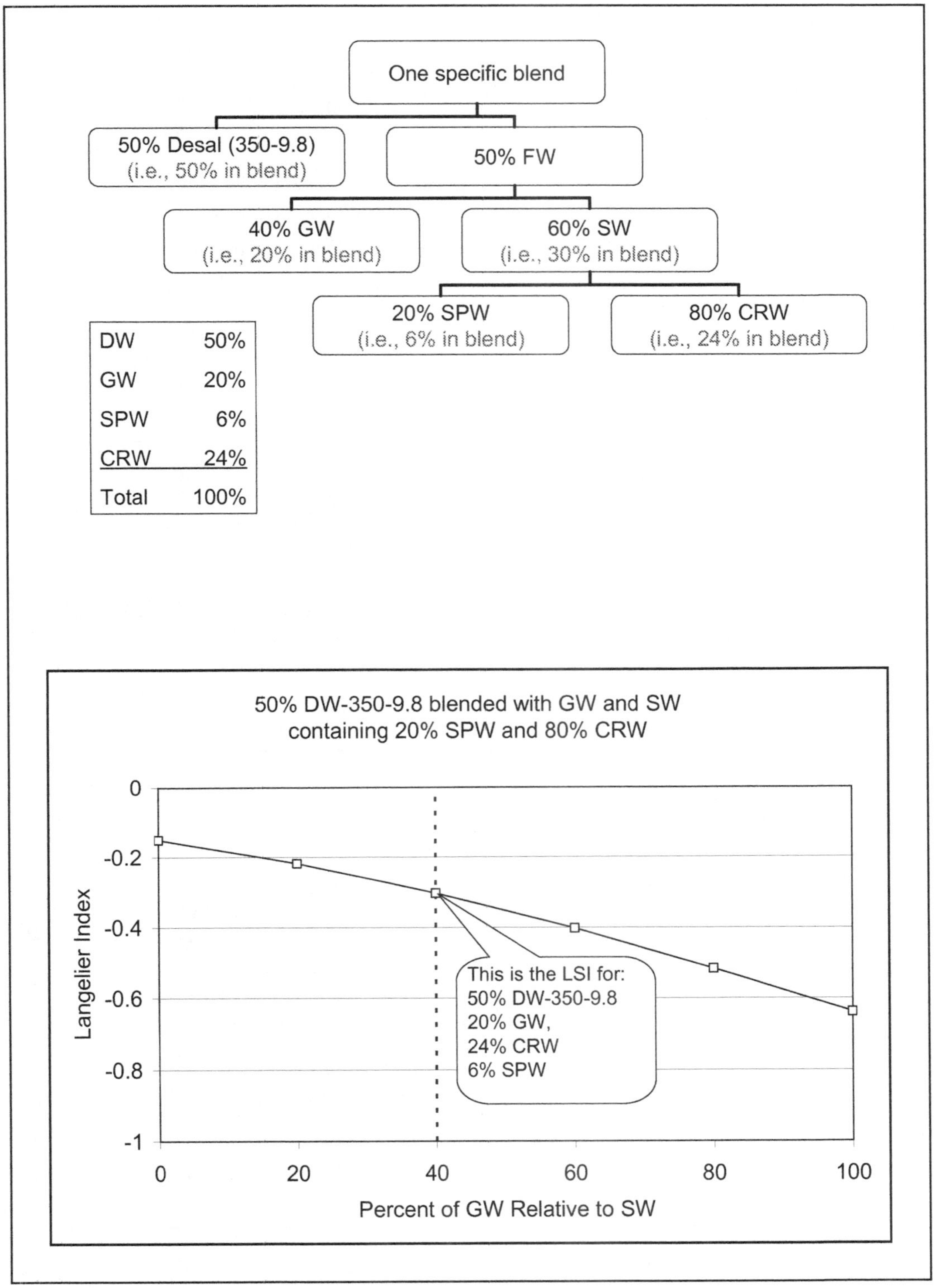

Figure 3.1 Example blend and corresponding LSI

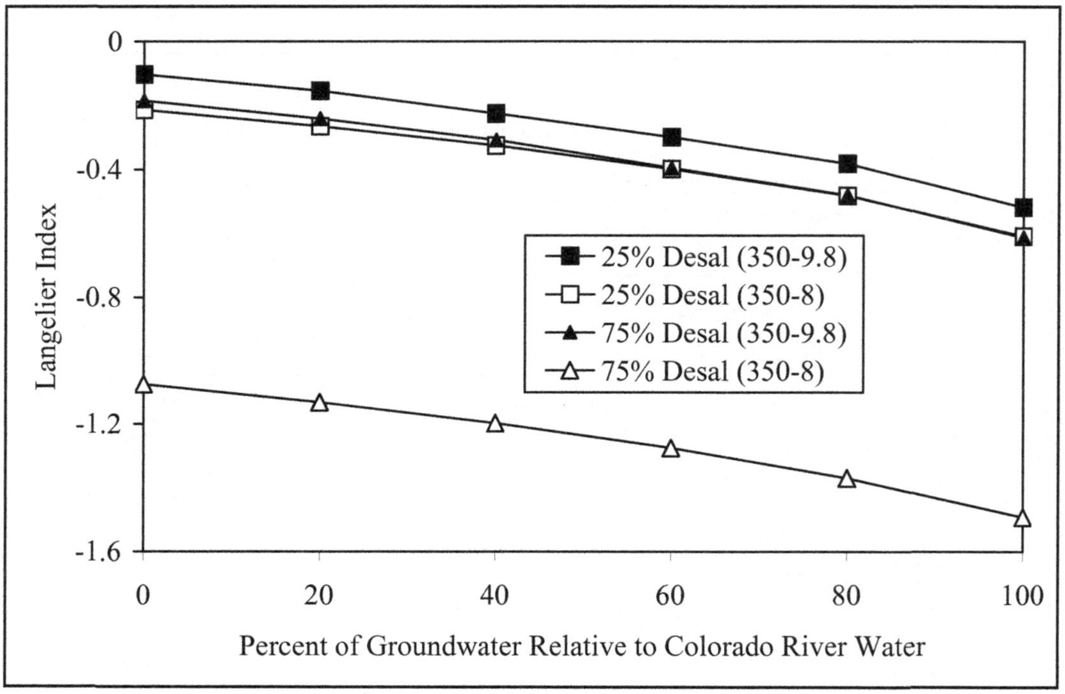

Figure 3.2 Effect of surface water composition on LSI for desalinated water (at 350 mg/L TDS and pH 8 or 9.8) blended with groundwater and Colorado River water

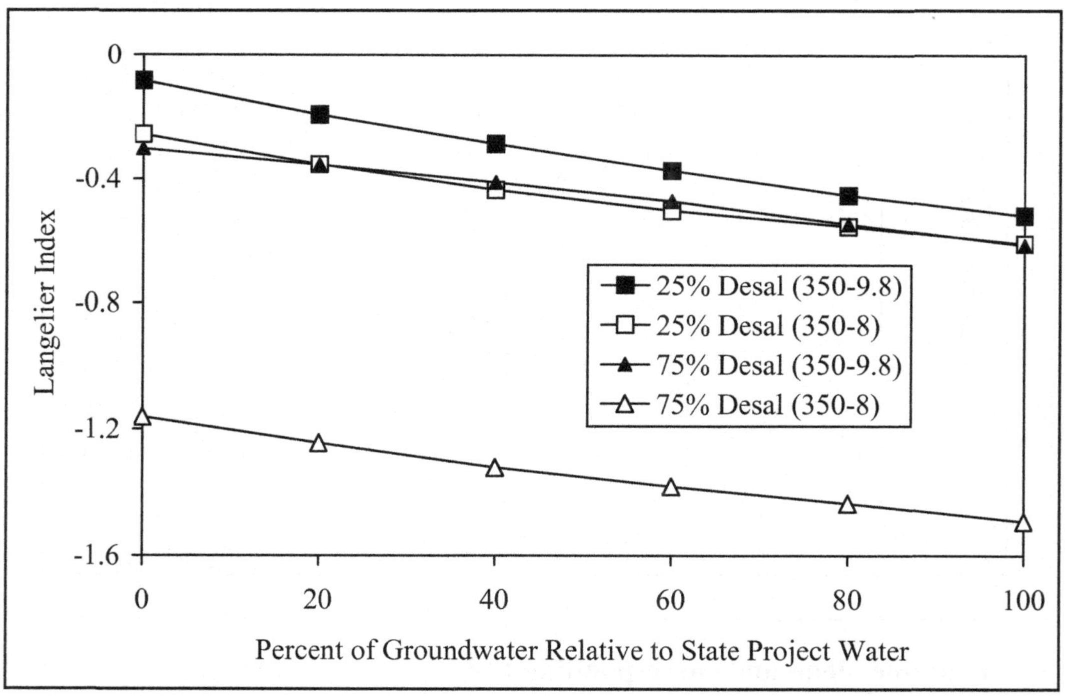

Figure 3.3 Effect of surface water composition on LSI for desalinated water (at 350 mg/L TDS and pH 8 or 9.8) blended with groundwater and State Project water

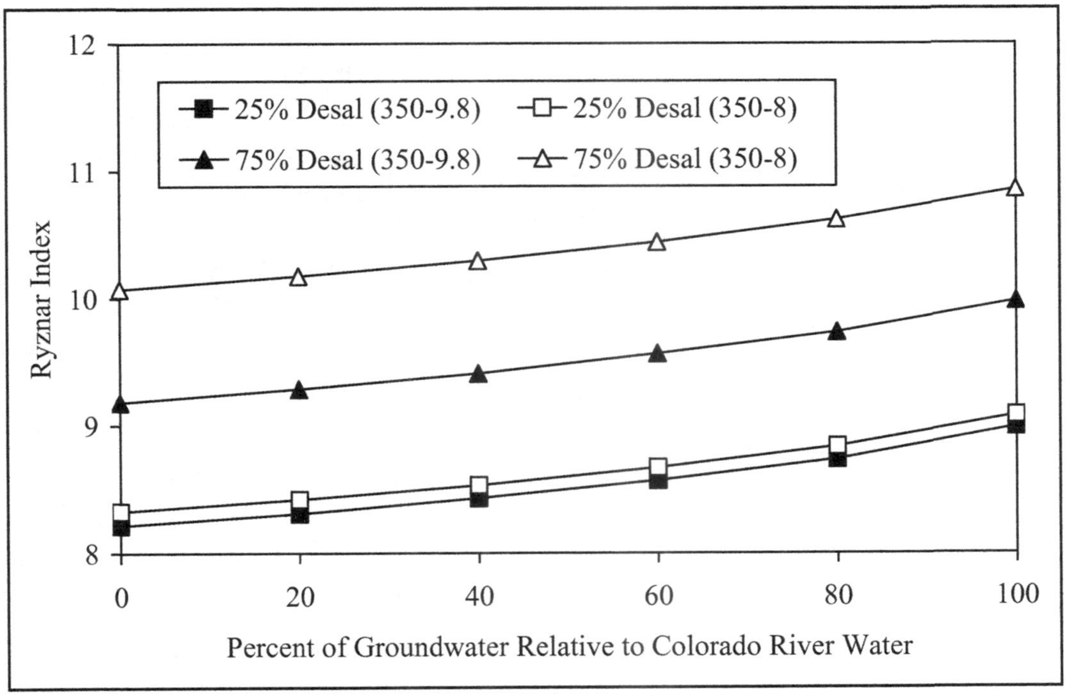

Figure 3.4 Effect of surface water composition on RSI for desalinated water (at 350 mg/L TDS and pH 8 or 9.8) blended with groundwater and Colorado River water

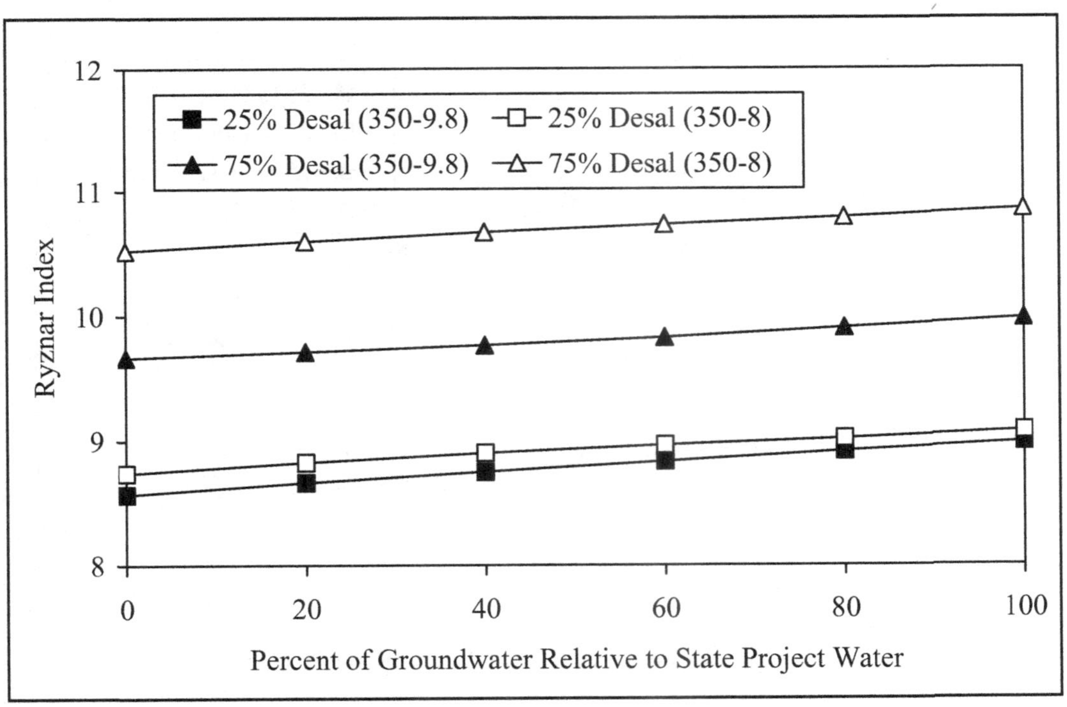

Figure 3.5 Effect of surface water composition on RSI for desalinated water (at 350 mg/L TDS and pH 8 or 9.8) blended with groundwater and State Project water

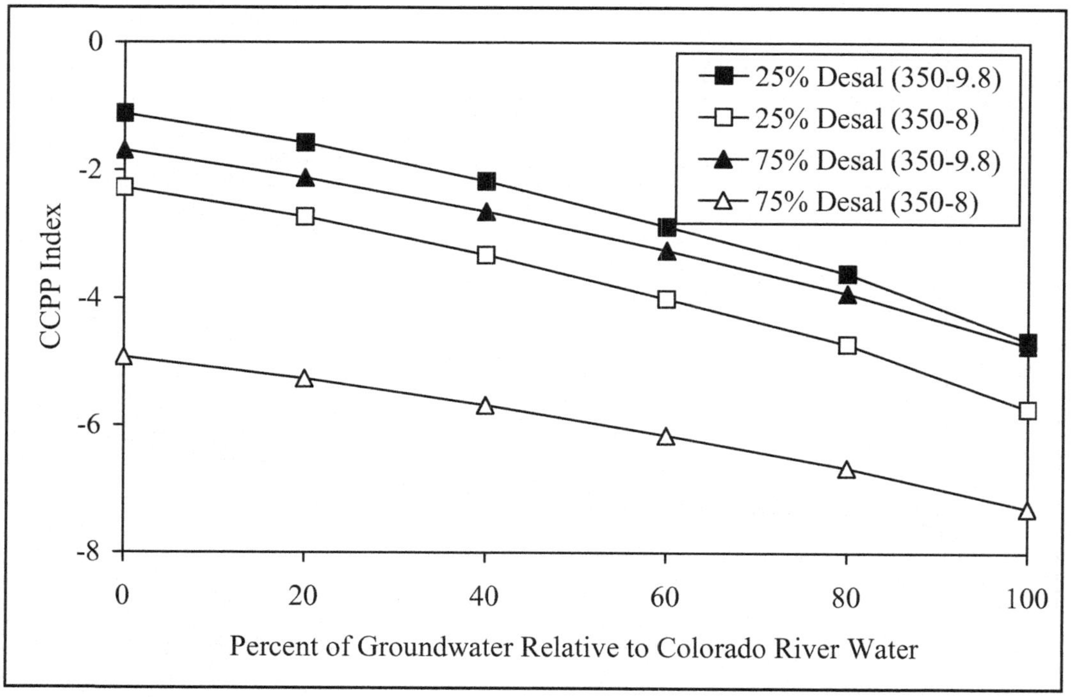

Figure 3.6 Effect of surface water composition on CCPP for desalinated water (at 350 mg/L TDS and pH 8 or 9.8) blended with groundwater and Colorado River water

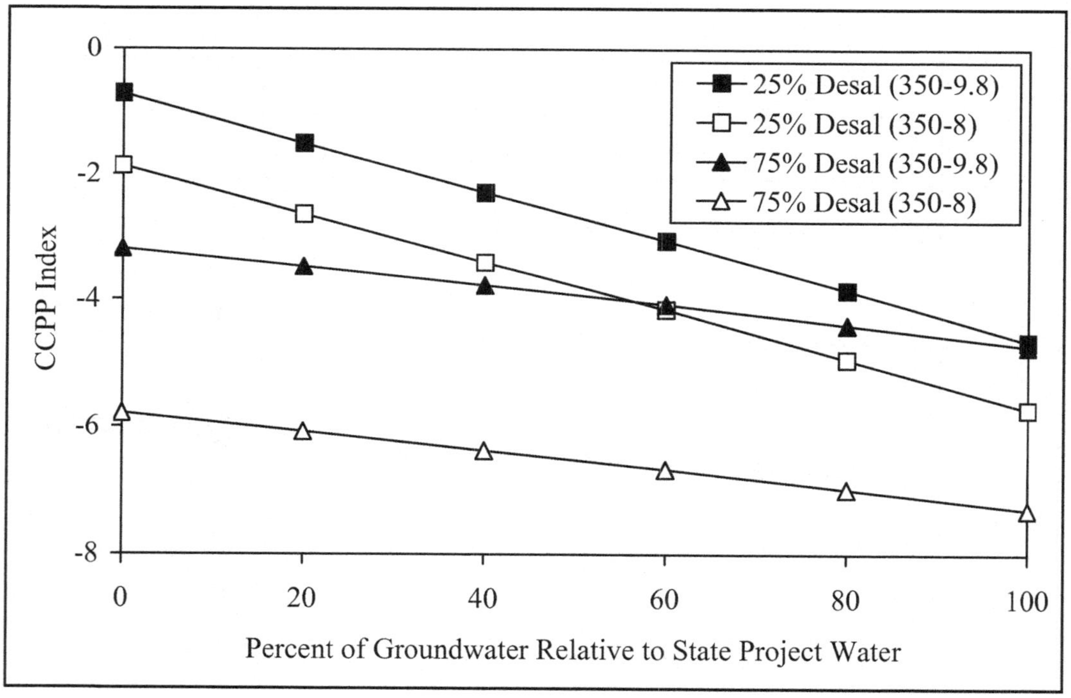

Figure 3.7 Effect of surface water composition on CCPP for desalinated water (at 350 mg/L TDS and pH 8 or 9.8) blended with groundwater and State Project water

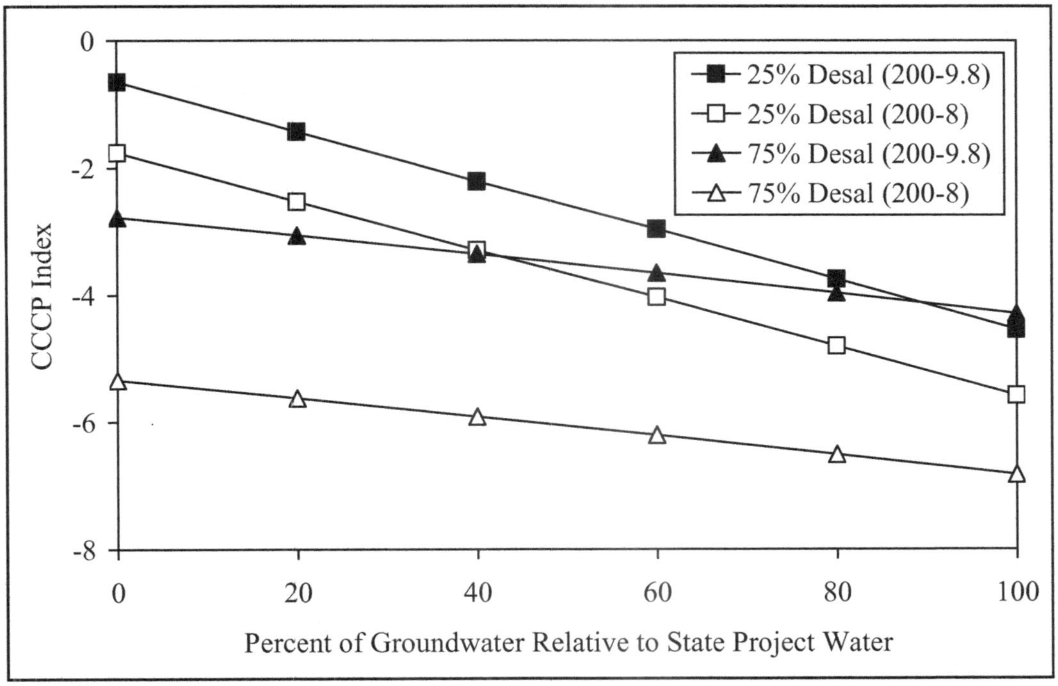

Figure 3.8 Effect of surface water composition on CCPP for desalinated water (at 200 mg/L TDS and pH 8 or 9.8) blended with groundwater and State Project water

Table 3.2 summarizes the results from Figures 3.2 through 3.8 by tabulating the values for which the blended water composition corresponds to a specific season. For the purposes of this report, LBWD operates its distribution system according to winter (October to April) and summer conditions (May to September). During winter conditions, the water that is served in the distribution system is virtually 100-percent surface water, and can either be 100 percent SPW or 100 percent CRW. During the summer, the water distributed is 100 percent groundwater.

For the summer condition, it appears that blending DW at pH 8.0 with GW would likely lead to corrosive conditions as indicated by the LSI and CCPP. In winter, if the proportion of DW at pH 8.0 is too high compared to SPW (e.g., 75 percent of the total blend) then corrosive conditions would prevail. Given these findings, LBWD has several options available to mitigate any corrosion potential from blending DW with existing finished water:

- Increase the corrosion inhibitor dosage (e.g., orthophosphate) in summer when GW is used, and in winter when SPW is used for SW
- Maintain the elevated pH needed for the boron removal step in DW (the impact of high pH on the chloramination of DW will need to be assessed)
- Maintain the proportion of DW in the final blend to 25 percent or less in winter if DW needs to be at pH 8.0

Table 3.2
Summary of corrosion index value for various blends of groundwater, surface water, and desalinated water

	25 percent DW* @ pH 8			25 percent DW* @ pH 9.8			75 percent DW* @ pH 8			75 percent DW* @ pH 9.8		
Corrosion	Summer	Winter		Summer	Winter		Summer	Winter		Summer	Winter	
Index	GW	SPW	CRW	GW	SPW	CRW	GW	SPW	CRW	GW	SPW	CRW
LSI	-0.6	-0.2	-0.2	-0.5	-0.1	-0.1	-1.5	-1.2	-1.1	-0.6	-0.2	-0.2
RSI	9.1	8.7	8.3	9.0	8.6	8.2	10.0	9.7	9.2	10.9	10.5	10.1
CCPP	-6.0	-2.0	-2.1	-5.0	-1.0	-1.2	-7.5	-6.0	-5.0	-5.0	-3.2	-1.8

* DW salinity target: 350 mg/L TDS
shaded cell indicates (mild) corrosive condition

CHAPTER 4
VIRAL CHALLENGES

BACKGROUND

The original scope of work included a viral challenge phase. That phase was intended to provide preliminary data on the inherent redundancy of LBWD's desalination system and the possible impact of recycling streams on virus accumulation. The main findings would have formed the basis of a more exhaustive virus challenge study that may be required by the California Department of Health Services (CDHS). MS-2 bacteriophage was the virus selected for the tests because it has become an accepted surrogate in the water industry.

System integrity towards viruses would have been compared with intact NF membranes and previously used NF membranes exhibiting 50-percent higher salt passage from exposure to chlorine. In one test, the damaged membranes would have been placed in the first stage, while in the subsequent test the damaged membrane would have been installed in the second stage. If time and budget allowed, an additional test would have used intact membranes with a cut O-ring because defective O-rings are reported as greater risk for viral passage. The MS-2 stock solution would have been continuously fed to the system influent line during each test.

LBWD staff conducted preliminary recovery tests of MS-2 coliphages (i.e., how many MS-2 phages could be enumerated after a few hours) prior to the challenge tests. Results are summarized in Table 4.1 and show non-detect concentrations for the male specific titer, and inconsistent recoveries for the somatic titer. As anticipated from literature surveys, when phages were inoculated in a water matrix with less than 500 mg/L TDS (i.e., Stage 2 permeate), recovery was high. Because the tests were conducted under lab conditions, which are more ideal than pilot plant conditions, the research team determined that running pilot viral challenge tests using MS-2 coliphages in seawater would not be meaningful, especially when log removal calculations involve phage quantification. This research phase was then changed to conduct a desktop investigation of possible alternative surrogates to MS-2 phages.

Table 4.1
MS-2 phage spiking recovery

Matrix		Concentration added (cfu/mL)	Male-specific titer (percent recovery)	Somatic titer (percent recovery)
Raw seawater				
	Test 1	100	ND	3.5
	Test 2	200	ND	24
	Test 3	200	ND	49
Stage 2 permeate		200		92

LITERATURE REVIEW

Marine Viruses

Recent advances in marine microbiology research have led to the discovery of numerous viruses in the sea. Many years of research and technological advances have determined that about 10 million viruses exist in 1 mL of seawater (Fuhrman 1999, Gantzer et al. 1998, Hughes 2003, Sinton et al. 1999). Over 140 of these viruses have been classified as pathogenic to humans ranging from less than 25 nm to about 2.8 μm (including their tail) (Bosh 1998, Danavaro et al. 2001). Their hosts may be found among all organisms that inhabit the sea, bacteria being the most abundant, and also phytoplankton, fish, and marine mammals. The most common types of enteric viruses are summarized in Table 4.2.

Table 4.2
Human enteric viruses that may be found in seawater

Genus	Popular name	Disease caused	Size (nm)
Enterovirus	Poliovirus	Paralysis, meningitis, fever	20 to 30
	Coxsackievirus, A,B	Herpangia, meningitis, fever, respiratory disease, hand-foot-and-mouth disease, myocarditis, heart anormalities, rush, pleurodynia, and possibly diabetes	
	Echovirus	Meningitis, fever, respiratory disease, rush, gastroenteritis	
Hepatovirus	Hepatitis A	Hepatitis	27 to 30
Reovirus	Human rotavirus	Gastroenteritis	70
Adenovirus	Human (mast)adenovirus	Gastroenteritis, respiratory disease, conjunctivitis	90 to 100
Calicivirus	Norovirus (Norwalk-like virus)	Gastroenteritis, fever	27 to 32
	Norwalk virus	Gastroenteritis, fever	
	Human calicivirus	Gastroenteritis	
	SRSV	Gastroenteritis	
Astrovirus	Human astrovirus	Gastroenteritis	28 to 30
Parvovirus	Human parvovirus	Gastroenteritis	18 to 26
Coronavirus	Human coronavirus	Gastroenteritis, respiratory disease	80 to 120
Torovirus	Human torovirus	Gastroenteritis	100 to 120

Table derived from data of Bosh (1998), Caballero et al. (2004), Duckmanton et al. (1998), Fuhrman and Schwalbach (2003), Garof et al. (1998), and Griffin et al. (2003)

Human enteric viruses enter the water environment through the discharge of sewage contaminated waters and runoff, including agricultural runoff as well as from feces from pelagic birds (Bosh 1998, Duckmanton et al. 1998, Griffin et al. 2003). The fate of microbial enteric pathogens may take many potential routes of transmission (rivers and lakes, ground water, water supplies, shellfish, recreational activities, etc) (Bosh 1998, Fuhrman 2001). Virus occurrences vary with location (Table 4.3) and temporal conditions (Callahan et al. 1995, Griffin et al. 2003).

Although viruses cannot replicate outside their host's tissue, they can survive for several weeks in marine waters. Once in the marine environment, viruses can survive in the water column or by attaching to particulate matter and accumulating in sediments (Callahan et al. 1995, Danavaro et al. 2001, Davies et al. 1995, Fuhrman 2001, Fuhrman and Schwalbach 2003, Gantzer et al. 1998). When pathogens are introduced to seawater, most are inactivated and broken down by a variety of natural processes, presumably of the same type that lead to the turnover of native marine bacteria, viruses, and protozoa (Hughes 2003).

Table 4.3
Occurrences of human viruses in marine environments

Virus	Location	Concentration/10 L
Enteroviruses	Galveston Bay, Texas	Data not available
	Texas Gulf Coast	0.1 - 10.5
	Waikiki and neighboring beaches, Hawaii	<0.1 - 130,000
	Sarasota estuary, Florida	17.0 - 77.0
	Charlotte Harbor estuary, Florida	100.0 – 200.0
	Santa Monica Bay, California	Presence/Absence only
Reoviruses	Beach along the Catania coastline of Sicily	Presence/Absence only
Reoviruses and enteroviruses	Fano Beach and Albani Channel Harbour, Italy	Presence/Absence only
Adenoviruses	Beaches from Malibu California to the Mexican border	8,800.0 – 75,000.0
Adenoviruses and enteroviruses	Beaches of Patras, Greece	1.2 - 6.0 (enteroviruses only)
Adenoviruses, enteroviruses, and HAV	Near-shore water of Barcelona, Spain	Presence/Absence only
Enteroviruses, HAV, and Norwalk virus	Canals and near-shore waters of Florida Keys	Presence/Absence only

Exposure to sunlight, higher temperatures and high microbial activity will shorten the survival of enteric viruses. The most important factor to the inactivation of viruses is temperature. According to some studies, tests have shown that in seawater, the time necessary to inactivate 90-percent of the virus is 671 days at 4°C and only 25 days at 25°C. Seawater salinity has little or virtually no effect on virus survival (Gantzer et al. 1998, Jiang et al. 2001). In the presence of suspended solids, 90 to 99 percent of the viruses are absorbed; this absorption does not provide much protection for viruses at low suspended solids concentration (from 3 to 15 mg/L), but a slight increase in viruses protection is noticed when higher levels of suspended solids are present (500 mg/L) (Gantzer et al. 1998).

Viruses attached to particulate matter (inorganic soil particles) are more likely to cause disease than free floating viruses. In a research in the Oder Lagoon (Southern Baltic) a model was used to predict the risk of infection from both unattached and particle-associated viruses. Researchers found that unattached viruses in rivers and at river mouths during the summer bathing season were not a risk. Viruses that were attached to particulate matter (suspended solids) extended their risk of infection from river mouths to coastal waters (Griffin et al. 2003).

Humans are exposed to enteric viruses in different manners ranging from oral ingestion to recreational contact (Bosh 1998). It has been determined that high counts of viruses are found in storm drains even during the relatively drier summer season. Overall, results confirm that viruses predominate near storm drains and after heavy rains. Illness can come from skin or wound infections, ear, sinus, and respiratory infections.

Marine Bacteria

Viruses are different from bacteria, in terms of size (bacteria are larger), chemical structure, and biological activity. Microbiological safety of beaches is determined by bacterial indicators, such as coliform or fecal coliform bacteria. The primary function of such monitoring is to detect beach contamination from sewage (Duckmanton et al. 1998, Jiang et al. 2001). Fecal indicator bacteria are natural inhabitants of the gastrointestinal tracts of humans and other warm-blooded animals and are released into the environment with feces (Davies et al. 1995, Jiang et al. 2001, Noble and Fuhrman 2000). Fecal bacteria generally do not survive in natural environments, because they have adapted to living in the gastrointestinal tract.

Sunlight, temperature, competition with bacteria found naturally in the water, predation by protozoa and other small organisms, and toxic industrial wastes are all believed to influence the survival of fecal indicator bacteria in the water. Studies have shown that fecal indicator bacteria survive from a few hours up to several days in seawater, but may survive for days or months in sediments, where they may be protected from sunlight and bacteriophages (Sinton et al. 1999). It is assumed that pathogens die at the same rate as fecal indicator bacteria.

The sources of fecal indicator bacteria include wastewater treatment plant effluents, combined sewer outfalls, drainage from septic tanks, runoff from agricultural fields, effluents from food processing plants (especially meats and beverages), and stormwater runoff (which carries animal and bird droppings). The likelihood that fecal indicator bacteria will survive for enumeration at a given water quality monitoring site is a function of the distance of the site from such sources, and also a function of the effect of all the environmental factors that influence bacterial survival.

Viral Surrogates

In most microbial assays for drinking water quality, the most common and recommended bacteriophage remains the MS-2 phage (Caballero et al. 2004), because it is a good surrogate in terms of microbial characteristics. However, MS-2 bacteriophage does not lend itself to good recoveries in a seawater matrix (sometimes less than 3 percent) because the high osmotic pressure results in lysing of the the phage. Most marine bacteria can be described as "hardy" and are able to withstand different environmental conditions, especially in marine waters. Table 4.4 describes the closely associated viruses found in common contaminated coastal waters. However, questions remain regarding the actual survival duration of these surrogate viruses in seawater and how complex and standardized are the analytical methods for their enumeration.

Regulatory Perspective

USEPA's 1991 Guidance Manual for Compliance with the Filtration and Disinfection Requirement for Public Water Systems Using Surface Water Sources (i.e., compliance with the Surface Water Treatment Rule – SWTR) states that:

> "Saline water sources such as the ocean are not generally considered to be subject to the requirements of the SWTR because of the low survival time of pathogens in a saline environment (…). Pathogens generally can only survive a few hours in saline water and any remaining pathogens should be removed or inactivated during desalination. However it is up to the Primacy Agency's discretion to determine which systems must meet the SWTR requirements." (p. 2-1)

Peer-reviewed articles confirm the relatively short persistence of various enteric and indicator viruses in seawater (Callahan et al. 1995) and therefore it is likely that survival of MS-2 phages in a seawater matrix would be poor, as demonstrated by the recovery test conducted earlier by LBWD. However, a large variety and quantity of indigenous viruses do exist in seawater and will require removal and/or inactivation.

Table 4.4
Possible surrogate animal viruses proposed by American Type Culture Collection (ATCC) for seawater-matrix tests

Classification	Agent	Original source	ATCC #
Coronavirus	Human enteric coronavirus	Stool from an infant with necrotizing enterocolitis	VR-1475
Adenovirus	Adenovirus type 9	Anal specimen from illness resembling rheumatoid arthritis	VR-10
Enterovirus	Echovirus 16	Fecal specimen from patient with aseptic meningitis	VR-1057
	Coxsackievirus B1	Rectal swap from man	VR-687
	Coxsackievirus B3	From man	VR-688

USEPA's SWTR Guidance Manual also states that:

> "(...) a membrane filtration method which is used for desalination (...) is effective for the removal of *Giardia* cysts and viruses and no demonstration is necessary *[i.e., provides at least 3-log removal for* Giardia *and 4-log removal for viruses]*" (p. 4-23)

However,

> "Although the SWTR requires a minimum of a 3-log removal/inactivation of *Giardia* cysts and a minum of 4-log removal/inactivation of viruses, it may be appropriate for the Primacy Agency to require greater removals/inactivations depending upon the degree of contamination within the source water." (p. 4-27)

> "The SWTR requires that disinfection be included as part of the treatment of surface water for potable use. (...) EPA recommends that (...) filtration systems provide sufficient disinfection to achieve a minimum of 0.5 log *Giardia* cyst and 2-log virus inactivation" (p. 4-25)

Although NF and RO are non-porous semi-permeable membranes that represent a barrier to viruses, the associated modules are generally not manufactured to be aseptic. It is clear from the last quote that a disinfection step should be incorporated in the future seawater desalination system, even if pathogen removal by the NF^2 process is greater than the minimal regulatory goal set forth in the SWTR, to keep the number of treatment barriers commensurate with the degree of contamination in the source water.

ALTERNATIVE SURROGATES

The size of viruses range from 30 to 300 nm, with the Norwalk virus, Hepatitis virus, calicivirus, poliovirus, coxsackievirus, and some enteroviruses in the low size range. The molecular weight of viruses is usually in the order of 10^5 to 10^7 amu. Therefore virus surrogates would require a size less than 30 nm and/or molecular weight less than 100,000 amu. Additionally, a conservative challenge particulate should have a neutral surface charge, because charged particles may interact with other particles and surfaces, thus enhancing removal. Finally, the surrogate should be cost effective, easy to use, and measurable. Four possible alternatives are evaluated below using the criteria for conservative surrogates.

Molecular Markers

A variety of macromolecular markers have been historically used to characterize the pore size or removal capabilities of membrane processes. For example, a polydisperse solution of polymers such as dextrans or polyethylene glycols (PEG) can determine the molecular weight cut off (MWCO) for UF membranes. Gel permeation chromatography (GPC) is often used to determine MWCO (at which 90 percent of the solute will be retained) because polymers with molecular weights between 1,000 to 10,000 amu are waxy. Other analytical instruments such as dynamic light scattering (DLS) can also be used to quantify the size, and molecular weight of macromolecular markers. However, GPC and DLS, and even more so TOC or conductivity monitoring instruments, are not sensitive enough to demonstrate more than 3-log removal, which would therefore require a high influent concentration that would irreversibly foul the membranes.

Starch

Starch has a molecular weight in the range of 10^5 to 10^6 amu and consists of two types of molecules, amylose (20 to 30 percent) and amylopectin (70 to 80 percent), both consisting of glucose polymers. Amylose molecules consist of single mostly unbranched chains of 500 to 20,000 glucose units, and hydrodynamic radius is 7 to 22 nm. Amylopectin is formed by branching of the amylose type glucose structure. Each amylopectin molecule contains up to 2,000,000 glucose residues in a compact structure with hydrodynamic radius of 21 to 75 nm (http://www.lsbu.ac.uk/water/hysta.html).

The mass, size, and quantity of starch can be measured by size exclusion chromatography (SEC) and DLS. Another property of starch however is its reaction with iodine (MW 100) to form a purple color. Based on the of the iodometric method (Standard Method 4500-Cl B), the purple color appears for a ratio of 1 g KI for 2.5 mg of starch. However, the latter method means the minimum starch concentration in the permeate should be 5 mg/L (when sample volume is 500 mL), which would require a starch concentration of at least 50 g/L in the raw water to show 4-log removal: such high starch concentration would likely cause irreversible fouling of the membranes.

Tracer Dyes

Fluorescent Rhodamine WT has been used to characterize RO and NF membranes. Rhodamine has high spectrophotometric absorbance (it absorbs green visible light at wavelength of 540 nm and emits an orange-red visible light at 575 nm) allowing detection levels in the ng/L, but can only be used with RO or tight NF membranes because the molecular weight is about 400 to 500 amu. Best practice recommends that the MWCO of a membrane be half the molecular weight of the solute to be retained. According to Dow Chemical and peer-reviewed analyses (Akbari et al. 2002) FilmTec's NF-90 membrane has a MWCO of 200 amu. Hence, Rhodamine could theoretically be used as a surrogate just based on its molecular size.

However, fouling occurs strongly on the membrane surface of polyamide membranes (Akbari et al. 2002), which is the active layer material of FilmTec's NF-90 membrane. Additionally, because the MWCO of Saehan NE-90 is unknown, using this surrogate method would be a one-time effort because the likelihood of irreversible fouling may be high. The fluorometric analysis of Rhodamine would require a fluorometer, a Rhodamine optical kit, cuvettes, secondary standards, and at least 1 gal of Rhodamine WT (20 percent aqueous solution).

Fluorescent Microspheres

Neutral, spherical-shaped, latex microspheres with a maximum diameter of 0.02 µm (20 nm) which are completely mono-dispersed may constitute a conservative surrogate to viruses. Nile red-fluorescent microspheres have broad excitation/emission bandwiths at 535/575 nm, which make them compatible with filter sets for Rhodamine. Such microspheres cost about $200 for 1.9 x 10^{12} microspheres/mL (= 0.2 g of microspheres in a 10 mL vial). Based on the detection level of fluorometer (~10^{-6} g/L Rhodamine), demonstrating a 4-log removal in a 2-hr test with a raw flow of about 50 L/min would require a concentration of 10 mg/L (assuming 100 percent recovery), which translates into 60 g of microspheres (300 vials) for the duration of the test (representing the treatment of about 6,000 L).

SUMMARY

Coastal waters with high population of marine birds and surrounded with land populated with animal feed operations will have higher concentrations of enteric viruses in the beach water columns, especially after rain events and at colder climates. In a study on viruses associated with sinking particles, despite high bacterial densities in the deepest sediment trap (400-m depth), no viruses were found there (Danavaro and Serresi 2000, Procotor and Fuhrman 1991). The lack of much higher virus/bacterium ratios in deep-sea sediments could be the result of less hospitable physical conditions that limit the ability of viruses to find new hosts.

Adsorption to settling particles and subsequent sedimentation has been suggested as one of the major mechanisms of virus removal from the water column. The highest virus densities observed in the top 1-cm portion of sediments were up to 3 orders of magnitude higher than those reported for the overlying water column (Procotor and Fuhrman 1991, Hara et al. 1996). Additionally, virus density (approximately 10^5 to 40×10^5 viruses/ml) decreased about 20-fold from the surface to a 5,000-m depth (Danavaro and Serresi 2000, Hara et al. 1996). Therefore any seawater intake should ideally be located away from the beach, be sufficiently deep below the water surface, and not disrupt sediment.

The proposed virus surrogate alternatives would either exceed the budget (fluorescent microspheres), or could irreversibly foul the membranes (molecular markers, starch, and tracer dye). Conducting a one-time test at a reasonable cost with Rhodamine WT (e.g., with pinched O-rings for some membrane modules in one or both stages) would have no statistical significance. The objective of the project was not to develop a new analytical method for viral challenge tests, but rather to apply an already verified method to test the system inherent redundancy. Given the above issues, as well as the time and budget constraints of the project, the challenge test was eliminated. The findings described in this section are important for the water industry and underscore the need for CDHS and other regulatory agencies to consider other surrogates than MS-2 phages for virus removal validation of a seawater desalination process.

CHAPTER 5
CONCLUSIONS AND RECOMMENDATIONS

CONCLUSIONS

The Long Beach method of using dual staged nanofiltration (NF^2) was shown to be able to desalt seawater to drinking water levels (i.e., less than 500 mg/L TDS) with 20 percent less energy than is theoretically needed for single-pass SWRO, and at a recovery close to 40 percent (typical SWRO system recovery is 50 percent). With the proper interstage pH adjustment (pH 10), it was also demonstrated that boron can be effectively reduced to less than California's boron Notification Level (1 mg/L). Although high pH is required to achieve sufficient boron removal in the second stage, the first-stage NF membranes tested removed a sufficient concentration of hardness ions such that the interstage pH adjustment is not expected to have any adverse effects on scaling for the second stage. The model developed in this report was able to predict permeate flow rate and water quality within 10 percent of pilot test results at various temperature, pressure, and raw flow, and it is anticipated that the model can be used to provide a guide to future testing at the prototype scale (300,000 gpd)

Suitable Membrane

The nature of the NF^2 process allows multiple membrane configurations to be used (e.g., different membrane selectivity in Stage 1 compared to Stage 2). Stage 2 was found to control the final salinity, and because the majority of the TDS in the second stage was in the form of NaCl, it was important to select an applicable NF membrane that rejected this salt well. From the optimization model, which was calibrated against the pilot-scale data, it appears that a membrane exhibiting 90 percent NaCl rejection (at 20°C and 35,000 mg/L TDS) would be suitable if the permeate flux is satisfactory (e.g., > 15 gfd).

A low salinity target (e.g., 200 mg/L) was shown to require more energy than a higher salinity goal (e.g., 350 mg/L). When compared to looser NF membranes, tighter NF membranes require less incremental change in the overall energy input as the final salinity target changes from 350 to 200 mg/L TDS. To compensate for the lower salt rejection, the looser membranes depend more on the recycle stream to dilute the feed water and thus achieve the final permeate salinity target, and therefore more flow needs to be pumped which requires greater energy. Membrane selectivity becomes essential in optimizing energy input: the greater membrane resistance exhibited by a tighter membrane requires increased operating pressure, while a looser membrane requires more flow to be processed to achieve the same water quality target through dilution of the feedwater. For tighter membranes, recycling flows to Stage 1 feed is of little value and may actually increase the energy required per volume of water treated.

Process Operation

To optimize energy consumption, the Stage 2 pressure should be set to maximize recovery to the point of not impacting final permeate quality. The overall system recovery is improved when Stage 1 pressure is maximized to the extent allowable by the manufacturers (typically 600 psi). Although both loose and tight NF membranes can desalt seawater to potable water quality, the additional energy required to dilute the feedwater by recycling process streams results in more energy that is required if a tighter membrane was used. Membrane selectivity therefore becomes essential as well in the trade off between recovery and energy required. Because Stage 1 requires 80 percent or more of the overall energy, any effort in saving energy should be focused on Stage 1. Because divalent ion removal principally occurs in the first stage, the potential for scaling is anticipated to be lessened in Stage 2 when pH is increased there for boron removal.

System Integration

Blending is needed to lessen the corrosivity of the desalinated water. It appears that of the corrosion indices considered CCPP shows the greatest sensitivity to the various water quality conditions. Several options are available to mitigate any corrosion potential from blending desalinated water with existing finished water:

- Increase corrosion inhibitor dosage as needed
- Maintain an elevated pH for the desalinated permeate (required for boron removal)
- Decrease proportion of desalinated water to a level where CCPP indicates the blend is not aggressive if pH from desalinated water needs to be at pH 8.0

The two-pass NF system will remove most of the seawater TOC so that formation of TTHM and HAA5 upon primary disinfection of the permeate with chlorine is unlikely. However, the permeate will include bromide ions, which will react with chlorine to form bromine and hypobromous acid (HOBr). Two issues are thus associated with bromide ions in the permeate:

- Bromide represents a source of chlorine demand, therefore the chlorine dose needs to be increased to meet disinfection requirements
- HOBr reacts with ammonia forming bromamines, which, unlike chloramines, will react with TOC when desalinated water is blended with surface water (brominated DBPs will form before reverting to bromide ions). Disinfectant residual depletion will slow down after all bromamines and HOBr have decomposed to bromide ions (chloramines do not oxidize bromide ions to bromine).

RECOMMENDATIONS

NF^2 Prototype Testing

Operating conditions suggested by the predictive model will guide the membrane selection and configuration of the long-term testing of the NF^2 process at a 300,000-gpd prototype plant, which will include a conventional SWRO train for side-by-side comparison. Other aspects to investigate will include a fouling potential assessment and further refinement in conserving the system energy (i.e., testing energy recovery devices on Stage 1 concentrate). The predictive model suggested that most of the desalting occurs within the first four elements in Stage 1 and the first three elements in Stage 2, and it is recommended that this be verified to further optimize the process.

Membrane Distribution

The relationship between recovery (water quantity) and rejection (water quality) was modeled to determine the benefit of mixing membranes with various salt rejection and flux characteristics on the overall system performance. Generally, the conditions that result in low overall energy required included using tighter membrane in Stage 1 (94 percent rejection) and looser membrane in Stage 2 (84 percent rejection). This suggested membrane distribution will need to be tested in the prototype plant

Critical Point

After a critical pressure was exceeded for some salts, rejection decreased in bench and pilot tests. Below this critical point, convection dominated and permeate water quality improved with increasing pressure. Once the critical point was reached, diffusion dominated and permeate water quality deteriorated with further increases in pressure. This phenomenon will be verified at the demonstration-scale to determine the balance of water quality and energy input.

Fouling Study

From SEM imaging and surface elemental analysis, it was determined that the majority of fouling can be attributed to a few localized deposits on the membrane surface as well as an organic coating over the entire surface. Elemental analyses revealed that the deposits consisted of diatoms (high in silica) as well as scale (e.g., Si, Fe, Al, and Ca). Long-term fouling investigations will need to be conducted at the prototype plant to understand the extent of such fouling potential and the corresponding cleaning techniques.

An increase in cross-flow velocity improved flux and rejection because the CP layer thickness was decreased. However, increasing cross-flow velocity may not be sufficient to overcome the increased particle loading to the membrane surface resulting from a high initial flux. Therefore, during prototype testing various loading rates on the membranes will be evaluated in their ability to mitigate fouling on the membrane surface. The loading rates can be varied by increasing raw water flow or modifying the number of parallel Stage 1 pressure vessels.

Disinfectant residual stability

Additional investigation is needed on residual disinfection to allow the process to be a viable alternative for a water resources portfolio. In particular, the impact of maintaining a high pH for corrosion control purposes on chloramination needs to be assessed. Controlling the effects of bromination (i.e., brominated residuals forming brominated DBPs and depleting disinfectant residual) will be essential for system implementation. A study could investigate several mitigation strategies such as "quenching" the oxidation demand from unreacted TOC of the surface waters with a chlorine boost (the additional formation TTHM and HAA5 could be balanced by the dilution factor associated with blending).

Corrosion Study

Because only a desktop corrosion study was performed, blending strategies and other related corrosion issues will need to be investigated in a year-long pipe-loop study (to cover the seasonal variations in water sources: groundwater in summer, surface water in winter), which will partly consist of actual piping samples from Long Beach's distribution system. Various permeate conditions (e.g., salinity, pH) resulting from prototype plant tests will be used to make up various blends.

Viral Surrogates

The poor recovery of MS-2 phage in seawater matrix demonstrates that research for suitable surrogates is needed for regulatory agencies concerned with virus removal validation of a seawater desalination process. Virus surrogates would require a size less than 30 nm and/or molecular weight less than 100,000 amu. Additionally, a conservative challenge particulate should have a neutral surface charge, because charged particles may interact with other particles and surfaces, thus enhancing removal. Finally, the surrogate should be cost effective, easy to use, and measurable. Some non-viral alternative surrogates were considered in this report in lieu of MS-2 bacteriophages, but each option had drawbacks to evaluate log removal (4-log removal is the minimum for virus removal per SWTR).

APPENDIX A
BENCH-SCALE DATA

Table A.1
Permeate water flux and ion rejection for NF-90 experiments with NaCl feed solution (I = 0.07M, 0.28M, 0.42M, 0.56M NaCl and T = 15°C and 20°C)

Conc. (M)	Conc. (g/L)	Temp. (°C)	Press. (psi)	Water flux (gfd)	Specific flux (gfd/psi)	R_{cation} (percent)	SP_{cation} (percent)	R_{anion} (percent)	SP_{anion} (percent)
0.07	4.09	15	100	23.3	0.23	90.9	9.1	93.3	6.7
			150	34.8	0.23	93.1	6.9	94.5	5.5
			200	42.2	0.21	93.7	6.3	95.5	4.5
			300	59.9	0.20	94.5	5.5	95.7	4.3
		20	100	25.0	0.25	89.5	10.5	92.3	7.7
			150	38.0	0.25	93.4	6.6	95.6	4.4
			200	49.1	0.25	93.3	6.7	95.8	4.2
			300	69.2	0.23	94.0	6.0	96.3	3.7
0.28	16.36	15	250	18.2	0.07	89.8	10.2	93.3	6.7
			300	26.8	0.09	92.5	7.5	95.6	4.4
			350	34.4	0.10	94.2	5.8	96.5	3.5
		20	250	18.9	0.08	88.9	11.1	93.2	6.8
			300	29.5	0.10	93.4	6.6	96.0	4.0
			350	38.0	0.11	94.6	5.4	96.7	3.3
0.42	24.55	15	350	17.3	0.05	92.3	7.7	95.4	4.6
			400	24.1	0.06	94.3	5.7	96.0	4.0
			450	30.9	0.07	95.1	4.9	96.8	3.2
		20	350	17.9	0.05	91.1	8.9	94.2	5.8
			400	26.8	0.07	93.5	6.5	96.0	4.0
			450	33.4	0.07	94.8	5.2	96.6	3.4
0.56	32.73	15	450	17.9	0.04	90.4	9.6	93.9	6.1
			500	22.8	0.05	91.4	8.6	93.7	6.3
			575	30.7	0.05	92.8	7.2	95.2	4.8
		20	450	18.7	0.04	88.4	11.6	92.5	7.5
			500	24.5	0.05	90.7	9.3	93.9	6.1
			575	32.9	0.06	92.3	7.7	95.1	4.9

Table A.2
Permeate water flux and ion rejection for NF-90 experiments with $CaCl_2$ feed solution
(I = 0.001M and 0.013M $CaCl_2$ and T = 15°C and 20°C)

Conc. (M)	Conc. (g/L)	Temp. (°C)	Press. (psi)	Water flux (gfd)	Specific flux (gfd/psi)	R_{cation} (percent)	SP_{cation} (percent)	R_{anion} (percent)	SP_{anion} (percent)
0.001	0.11	15	150	35.1	0.23	99.0	1.0	96.7	3.3
			225	53.2	0.24	99.2	0.8	97.6	2.4
			300	68.7	0.23	99.3	0.7	97.9	2.1
		20	150	39.8	0.27	98.9	1.1	96.9	3.1
			225	68.7	0.31	99.2	0.8	97.7	2.3
			300	79.2	0.26	99.3	0.7	97.9	2.1
0.013	1.44	15	165	32.6	0.20	99.7	0.3	98.1	1.9
			500	99.7	0.20	99.8	0.2	98.8	1.2
			575	110.0	0.19	99.8	0.2	98.9	1.1
		20	165	34.5	0.21	99.7	0.3	97.8	2.2
			500	110.1	0.22	99.8	0.2	98.8	1.2
			575	123.2	0.21	99.8	0.2	98.8	1.2

Table A.3
Permeate water flux and ion rejection for NF-90 experiments with $MgCl_2$ feed solution
(I = 0.003M and 0.029M $MgCl_2$ and T = 15°C and 20°C)

Conc. (M)	Conc. (g/L)	Temp. (°C)	Press. (psi)	Water flux (gfd)	Specific flux (gfd/psi)	R_{cation} (percent)	SP_{cation} (percent)	R_{anion} (percent)	SP_{anion} (percent)
0.003	0.29	15	145	36.4	0.25	98.6	1.4	95.3	4.7
			225	55.7	0.25	98.7	1.3	95.8	4.2
			300	74.2	0.25	99.2	0.8	96.8	3.2
		20	145	40.7	0.28	99.1	0.9	96.8	3.2
			225	65.9	0.29	99.4	0.6	97.3	2.7
			300	88.0	0.29	99.3	0.7	97.3	2.7
0.029	2.76	15	180	35.9	0.20	99.6	0.4	98.9	1.1
			500	99.7	0.20	99.4	0.6	98.8	1.2
			575	110.6	0.19	99.8	0.2	99.2	0.8
		20	180	38.9	0.22	99.7	0.3	99.5	0.5
			500	116.4	0.23	99.5	0.5	98.6	1.4
			575	126.3	0.22	99.7	0.3	99.3	0.7

Table A.4
Permeate water flux and ion rejection for NF-90 experiments with $MgSO_4$ feed solution
(I = 0.003M and 0.029M $MgSO_4$ and T = 15°C and 20°C)

Conc. (M)	Conc. (g/L)	Temp. (°C)	Press. (psi)	Water flux (gfd)	Specific flux (gfd/psi)	R_{cation} (percent)	SP_{cation} (percent)	R_{anion} (percent)	SP_{anion} (percent)
0.003	0.36	15	150	31.5	0.21	99.8	0.2	99.8	0.2
			225	45.7	0.20	99.8	0.2	99.8	0.2
			300	59.8	0.20	99.8	0.2	99.8	0.2
		20	150	34.4	0.23	99.8	0.2	99.8	0.2
			225	52.7	0.23	99.9	0.1	99.8	0.2
			300	68.9	0.23	99.8	0.2	99.8	0.2
0.031	3.73	15	170	27.4	0.16	99.9	0.1	99.8	0.2
			500	73.9	0.15	99.7	0.3	99.6	0.4
			575	81.4	0.14	99.7	0.3	99.5	0.5
		20	170	28.2	0.17	99.9	0.1	99.8	0.2
			500	86.6	0.17	99.8	0.2	99.6	0.4
			575	93.6	0.16	99.6	0.4	99.5	0.5

Table A.5
Permeate water flux and ion rejection for NF-90 experiments with KCl feed solution
(I = 0.01M KCl and T = 15°C and 20°C)

Conc. (M)	Conc. (g/L)	Temp. (°C)	Press. (psi)	Water flux (gfd)	Specific flux (gfd/psi)	R_{cation} (percent)	SP_{cation} (percent)	R_{anion} (percent)	SP_{anion} (percent)
0.01	0.75	15	150	30.9	0.21	91.4	8.6	89.9	10.1
			500	98.9	0.20	94.9	5.1	95.3	4.67
			575	112.5	0.20	94.6	5.4	95.0	4.97
		20	150	32.3	0.22	92.3	7.7	92.1	7.9
			500	115.1	0.23	95.2	4.8	95.7	4.3
			575	125.6	0.22	94.9	5.1	95.4	4.6

Table A.6
Permeate water flux and ion rejection for NF-90 experiments with NaBr feed solution (I = 0.01M NaBr and T = 15°C and 20°C)

Conc. (M)	Conc. (g/L)	Temp. (°C)	Press. (psi)	Water flux (gfd)	Specific flux (gfd/psi)	R_{cation} (percent)	SP_{cation} (percent)	R_{anion} (percent)	SP_{anion} (percent)
0.001	0.10	15	150	29.5	0.20	91.1	8.9	94.6	5.4
			500	96.1	0.19	96.2	3.8	95.9	4.1
			575	107.4	0.19	95.6	4.4	95.5	4.5
		20	150	31.2	0.21	94.1	5.9	94.4	5.6
			500	106.9	0.21	96.4	3.6	96.2	3.8
			575	120.0	0.21	95.0	5.0	95.8	4.2

Table A.7
Permeate water flux and ion rejection for NF-90 experiments with $NaHCO_3$ feed solution (I = 0.003M $NaHCO_3$ and T = 15°C and 20°C)

Conc (M)	Conc (g/L)	Temp (°C)	Press. (psi)	Water flux (gfd)	Specific flux (gfd/psi)	R_{cation} (percent)	SP_{cation} (percent)	R_{anion} (percent)	SP_{anion} (percent)
0.003	0.25	15	150	25.9	0.17	99.0	1.0	99.3	0.7
			500	83.2	0.17	98.5	1.5	98.8	1.2
			575	93.0	0.16	98.0	2.0	98.5	1.5
		20	150	27.4	0.18	99.2	0.8	99.5	0.5
			500	96.9	0.19	99.1	0.9	99.4	0.6
			575	106.7	0.19	98.6	1.4	99.1	0.9

Table A.8
Permeate water flux and ion rejection for NF-90 experiments with KI feed solution (I = 6.0×10^{-6} M KI and T = 15°C and 20°C)

Conc. (M)	Conc. (mg/L)	Temperature (°C)	Pressure (psi)	Water flux (gfd)	Specific flux (gfd/psi)	R_{anion} (percent)	SP_{aion} (percent)
6.0×1 0^{-6}	1	15	150	33.5	0.22	84.1	15.9
			300	63.8	0.21	90.7	9.3
			575	114.5	0.20	88.9	11.1
		20	150	36.5	0.24	85.7	14.3
			300	73.7	0.25	89.3	10.7
			575	133.4	0.23	89.2	10.8

Table A.9
Permeate water flux and ion rejection for NF-90 experiments with H_3BO_3 feed solution
($I = 4.6\times10^{-4}$ M Boron and T = 15°C and 20°C)

Conc. (M)	Conc. (mg/L)	pH	Temp (°C)	Pressure (psi)	Water flux (gfd)	Specific flux (gfd/psi)	R_{cation} (percent)	SP_{cation} (percent)
4.6×10^{-4}	5	7.8	15	150	24.5	0.16	57.1	42.9
				300	46.9	0.16	66.2	33.8
				575	88.1	0.15	64.5	35.5
			20	150	26.5	0.18	49.9	50.1
				300	54.8	0.18	60.3	39.7
				575	102.9	0.18	59.9	40.1
		9.8	15	150	23.1	0.15	91.0	9.0
				300	45.5	0.15	94.2	5.8
				575	85.5	0.15	93.0	7.0
			20	150	26.6	0.18	85.1	14.9
				300	53.0	0.18	89.1	10.9
				575	99.4	0.17	87.8	12.2

Table A.10
Permeate water flux and ion rejection for NE-90 experiments with NaCl feed solution (I = 0.07M, 0.28M, 0.42M, 0.56M NaCl and T = 15°C and 20°C)

Conc. (M)	Conc. (g/L)	Temp. (°C)	Pressure (psi)	Water flux (gfd)	Specific flux (gfd/psi)	R_{cation} (percent)	SP_{cation} (percent)	R_{anion} (percent)	SP_{anion} (percent)
0.07	4.09	15	100	12.05	0.12	89.14	10.86	86.49	13.51
			150	32.27	0.16	94.27	5.73	93.08	6.92
			200	52.23	0.17	95.63	4.37	93.92	6.08
			300	11.49	0.11	87.82	12.18	85.93	14.07
		20	100	38.78	0.19	94.31	5.69	93.08	6.92
			150	62.63	0.21	95.48	4.52	94.33	5.67
			200	14.09	0.06	88.04	11.96	86.15	13.85
			300	22.82	0.08	91.16	8.84	89.64	10.36
0.28	16.36	15	250	28.10	0.08	92.65	7.35	91.16	8.84
			300	14.43	0.06	86.79	13.21	84.84	15.16
			350	24.44	0.08	91.15	8.85	89.79	10.21
		20	250	32.96	0.09	92.65	7.35	91.41	8.59
			300	16.25	0.05	88.80	11.20	86.15	13.85
			350	22.55	0.06	91.29	8.71	89.56	10.44
0.42	24.55	15	350	28.20	0.06	92.47	7.53	91.11	8.89
			400	16.74	0.05	87.03	12.97	84.87	15.13
			450	25.33	0.06	90.60	9.40	89.09	10.91
		20	350	31.61	0.07	91.93	8.07	90.37	9.63
			400	16.66	0.04	87.62	12.38	85.19	14.81
			450	21.08	0.04	90.03	9.97	88.30	11.70
0.56	32.73	15	450	28.86	0.05	91.98	8.02	90.45	9.55
			500	18.68	0.04	86.89	13.11	84.26	15.74
			575	23.41	0.05	89.08	10.92	86.48	13.52
		20	450	31.98	0.06	91.33	8.67	89.35	10.65
			500	12.05	0.12	89.14	10.86	86.49	13.51
			575	32.27	0.16	94.27	5.73	93.08	6.92

Table A.11
Permeate water flux and ion rejection for NE-90 experiments with $CaCl_2$ feed solution (I = 0.001M and 0.013M $CaCl_2$ and T = 15°C and 20°C)

Conc. (M)	Conc. (g/L)	Temp. (°C)	Pressure (psi)	Water flux (gfd)	Specific flux (gfd/psi)	R_{cation} (percent)	SP_{cation} (percent)	R_{anion} (percent)	SP_{anion} (percent)
0.001	0.11	15	150	33.5	0.22	92.9	7.1	92.2	7.8
			225	48.7	0.22	95.3	4.7	95.0	5.0
			300	63.7	0.21	95.9	4.1	95.3	4.7
		20	150	37.2	0.25	93.5	6.5	93.7	6.3
			225	56.2	0.25	95.6	4.4	95.1	4.9
			300	74.4	0.25	96.1	3.9	95.5	4.5
0.013	1.44	15	165	33.2	0.20	93.7	6.3	94.3	5.7
			500	97.8	0.20	95.6	4.4	95.9	4.1
			575	107.4	0.19	95.9	4.1	96.1	3.9
		20	165	35.7	0.22	94.2	5.8	94.6	5.4
			500	114.0	0.23	95.9	4.1	96.1	3.9
			575	127.4	0.22	95.9	4.1	96.1	3.9

Table A.12
Permeate water flux and ion rejection for NE-90 experiments with $MgCl_2$ feed solution (I = 0.003M and 0.029M $MgCl_2$ and T = 15°C and 20°C)

Conc. (M)	Conc. (g/L)	Temp. (°C)	Pressure (psi)	Water flux (gfd)	Specific flux (gfd/psi)	R_{cation} (percent)	SP_{cation} (percent)	R_{anion} (percent)	SP_{anion} (percent)
0.003	0.29	15	145	33.6	0.22	92.7	7.3	92.5	7.5
			225	49.5	0.22	94.6	5.4	93.9	6.1
			300	68.2	0.23	95.2	4.8	94.5	5.5
		20	145	36.3	0.24	93.7	6.3	92.7	7.3
			225	58.2	0.26	94.9	5.1	94.2	5.8
			300	79.0	0.26	95.3	4.7	94.4	5.6
0.029	2.76	15	180	34.9	0.21	96.3	3.7	95.6	4.4
			500	95.1	0.19	97.1	2.9	96.7	3.3
			575	104.3	0.18	97.0	3.0	96.8	3.2
		20	180	39.1	0.23	96.2	3.8	95.6	4.4
			500	108.8	0.22	97.8	2.2	97.0	3.0
			575	121.3	0.21	98.0	2.0	97.1	2.9

Table A.13
Permeate water flux and ion rejection for NE-90 experiments with $MgSO_4$ feed solution
(I = 0.003M and 0.029M $MgSO_4$ and T = 15°C and 20°C)

Conc. (M)	Conc. (g/L)	Temp. (°C)	Pressure (psi)	Water flux (gfd)	Specific flux (gfd/psi)	R_{cation} (percent)	SP_{cation} (percent)	R_{anion} (percent)	SP_{anion} (percent)
0.003	0.36	15	150	35.8	0.24	95.3	4.7	94.6	5.4
			225	53.1	0.24	96.3	3.7	95.6	4.4
			300	69.5	0.23	96.9	3.1	95.8	4.2
		20	150	40.6	0.27	96.6	3.4	96.3	3.7
			225	62.2	0.28	97.2	2.8	96.6	3.4
			300	80.9	0.27	97.5	2.5	97.3	2.7
0.031	3.73	15	170	31.1	0.18	97.6	2.4	97.1	2.9
			500	93.8	0.19	97.6	2.4	97.1	2.9
			575	103.2	0.18	97.3	2.7	96.7	3.3
		20	170	37.7	0.22	97.0	3.0	96.3	3.7
			500	111.5	0.22	97.6	2.4	97.1	2.9
			575	120.5	0.21	97.8	2.2	97.4	2.6

Table A.14
Permeate water flux and ion rejection for NE-90 experiments with KCl feed solution
(I = 0.01M KCl and T = 15°C and 20°C)

Conc. (M)	Conc. (g/L)	Temp. (°C)	Pressure (psi)	Water flux (gfd)	Specific flux (gfd/psi)	R_{cation} (percent)	SP_{cation} (percent)	R_{anion} (percent)	SP_{anion} (percent)
0.01	0.75	15	150	30.2	0.20	94.3	5.7	93.4	6.6
			500	85.8	0.17	95.7	4.3	94.7	5.3
			575	95.4	0.17	95.8	4.2	94.4	5.6
		20	150	28.9	0.19	94.2	5.8	93.0	7.0
			500	102.3	0.20	95.9	4.1	94.3	5.7
			575	113.6	0.20	96.1	3.9	95.0	5.0

Table A.15
Permeate water flux and ion rejection for NE-90 experiments with NaBr feed solution (I = 0.01M NaBr and T = 15°C and 20°C)

Conc. (M)	Conc. (g/L)	Temp. (°C)	Pressure (psi)	Water flux (gfd)	Specific flux (gfd/psi)	R_{cation} (percent)	SP_{cation} (percent)	R_{anion} (percent)	SP_{anion} (percent)
0.001	0.10	15	150	31.1	0.21	82.7	17.3	95.2	4.8
			500	96.0	0.19	95.0	5.0	96.0	4.0
			575	109.0	0.19	96.2	3.8	96.8	3.2
		20	150	33.6	0.22	94.7	5.3	94.1	5.9
			500	115.9	0.23	96.5	3.5	96.5	3.5
			575	127.4	0.22	96.5	3.5	95.9	4.1

Table A.16
Permeate water flux and ion rejection for NE-90 experiments with $NaHCO_3$ feed solution (I = 0.003M $NaHCO_3$ and T = 15°C and 20°C)

Conc. (M)	Conc. (g/L)	Temp. (°C)	Pressure (psi)	Water flux (gfd)	Specific flux (gfd/psi)	R_{cation} (percent)	SP_{cation} (percent)	R_{anion} (percent)	SP_{anion} (percent)
0.003	0.25	15	150	28.6	0.19	98.0	2.0	99.1	0.9
			500	89.3	0.18	98.4	1.6	99.3	0.7
			575	101.3	0.18	98.4	1.6	99.4	0.6
		20	150	31.1	0.21	97.7	2.3	98.6	1.4
			500	106.3	0.21	98.3	1.7	99.1	0.9
			575	120.9	0.21	98.2	1.8	99.1	0.9

Table A.17
Permeate water flux and ion rejection for NE-90 experiments with KI feed solution (I = 6.0×10^{-6} M KI and T = 15°C and 20°C)

Conc. (M)	Conc. (mg/L)	Temp. (°C)	Pressure (psi)	Water flux (gfd)	Specific flux (gfd/psi)	R_{anion} (percent)	SP_{anion} (percent)
6.0×10^{-6}	1	15	150	28.2	0.19	89.2	10.8
			300	55.5	0.18	91.1	8.9
			575	102.1	0.18	91.2	8.8
		20	150	32.1	0.21	90.3	9.7
			300	65.5	0.22	91.0	9.0
			575	117.5	0.20	91.1	8.9

Table A.18
Permeate water flux and ion rejection for NE-90 experiments with H_3BO_3 feed solution
(I = 4.6×10^{-4} M Boron and T = 15°C and 20°C)

Conc. (M)	Conc. (mg/L)	pH	Temp. (°C)	Pressure (psi)	Water flux (gfd)	Specific flux (gfd/psi)	R_{cation} (percent)	SP_{cation} (percent)
4.6×10^{-4}	5	7.8	15	150	32.5	0.22	65.9	34.1
				300	60.4	0.20	78.6	21.4
				575	110.9	0.19	79.1	20.9
			20	150	34.5	0.23	58.8	41.2
				300	72.5	0.24	72.9	27.1
				575	134.6	0.23	74.6	25.4
		9.8	15	150	34.8	0.23	81.7	18.3
				300	67.5	0.23	90.3	9.7
				575	123.3	0.21	89.1	10.9
			20	150	40.2	0.27	75.0	25.0
				300	81.7	0.27	87.6	12.4
				575	145.8	0.25	85.2	14.8

APPENDIX B
SELECTED PILOT DATA FOR FILMTEC NF-90 AND SAEHAN NE-90

Table B.1
Pilot test results for FilmTec NF-90 at 520 psi in Stage 1, target salinity of 200 mg/L in Stage 2 permeate, and no Stage 1 permeate return.

5/10/2004	Water location						
Test 1	Raw	1F	1P/2F	1R	1C	2P	2C
Temperature (° F)	64.0	64.0	68.9		71.0	72.1	70.9
Pressure (psi)	100	520	170		510		150
Flow (gpm)	12.9	16.0	6.8		9.1	3.8	3.1
pH	7.59	7.64	7.64		7.91	7.30	7.47
Alkalinity (mg/L as $CaCO_3$)	117.3	97.6	8.4		164.7	3.2	13.1
Boron (mg/L)	4.8	4.7	3.3		6.0	2.3	4.4
Bromide (mg/L)	55.6	49.1	7.5		67.0	0.5	13.3
Calcium (mg/L)	419	335.0	7.3		585.0	0.6	15.0
Chloride (mg/L)	18939	15641	1985		25063	109	3991
Fluoride (mg/L)	0.63	0.59	0.10		0.68	0.10	0.10
Iodide (μg/L)							
Potassium (mg/L)	362	310	48		518	2.25	98
Magnesium (mg/L)	861	548.0	16.8		998.0	4.4	47.0
Sodium (mg/L)	10156	8903	1251		14428	71	2615
Silica (mg/L)	1.03	0.88	0.10		1.48	0.10	0.1
Sulfate (mg/L)	2560	2232	27		3730	0.1	59
TDS (mg/L)	37099	31770	3514		53574	206	7928
Langelier Saturation Index	0.40	0.28	-2.30		1.07	-4.0	-1.98
Recovery (percent)	29		43			55	
Flux (gfd)			6.0			16.4	
Energy (kWh/KGal)						15.9	

> For column headings, the preceding digit denotes Stage 1 or 2, F: feed, P permeate, C: concentrate, R: return
> Pilot unit configuration: 4 vessels/5 elements in Stage 1 and 1 vessel/4 elements in Stage 2 (element area: 82 ft^2)
> Energy based on Eq. (2.1)

Table B.2
Pilot test results for FilmTec NF-90 at 520 psi in Stage 1, target salinity of 350 mg/L in Stage 2 permeate, and no Stage 1 permeate return.

5/10/2004	Water location						
Test 2	Raw	1F	1P/2F	1R	1C	2P	2C
Temperature (° F)	67.3	68.54	70.7		72.7	73.9	74.8
Pressure (psi)		525	220		510		210
Flow (gpm)	14.2	15.9	6.0		9.8	4.4	1.7
pH	7.59	7.70	7.61			7.25	
Alkalinity (mg/L as $CaCO_3$)	117.3	107.3	9.4			3.3	
Boron (mg/L)	4.8	5.0	3.4			2.7	
Bromide (mg/L)	55.6	54.1	8.6			1.0	
Calcium (mg/L)	419	373.0	8.1			0.6	
Chloride (mg/L)	18939	17494	2125			206	
Fluoride (mg/L)	0.63	0.61	0.10			0.10	
Iodide (µg/L)							
Potassium (mg/L)	362	344	55			4.34	
Magnesium (mg/L)	861	693.0	18.3			5.2	
Sodium (mg/L)	10156	9766	1455			134	
Silica (mg/L)	1.03	1.02	0.10			0.10	
Sulfate (mg/L)	2560	2500	31			1.1	
TDS (mg/L)	35963	34478	4142		53567	380	13492
Langelier Saturation Index	0.43	0.47	-2.22			-4.04	
Recovery (percent)	31		38			72	
Flux (gfd)			5.3			19.0	
Energy (kWh/KGal)						13.8	

> For column headings, the preceding digit denotes Stage 1 or 2, F: feed, P permeate, C: concentrate, R: return
> Pilot unit configuration: 4 vessels/5 elements in Stage 1 and 1 vessel/4 elements in Stage 2 (element area: 82 ft^2)
> Energy based on Eq. (2.1)

Table B.3
Pilot test results for FilmTec NF-90 at 580 psi in Stage 1, target salinity of 350 mg/L in Stage 2 permeate, and no Stage 1 permeate return.

5/11/2004	Water location						
Test 3	Raw	1F	1P/2F	1R	1C	2P	2C
Temperature (° F)	64.4	68.36	70.9		72.0	73.4	73.2
Pressure (psi)		579	210		571		198
Flow (gpm)	10.2	12.6	6.6		6.0	4.2	2.4
pH	7.98	7.92	7.51		7.79	7.65	7.39
Alkalinity (mg/L as $CaCO_3$)	117.3	95.8	16.3			3.2	
Boron (mg/L)	4.8	4.9	3.5			2.5	
Bromide (mg/L)	55.6	47.8	10.8			0.8	
Calcium (mg/L)	419	332.0	9.1			0.6	
Chloride (mg/L)	18939	15340	2328			193	
Fluoride (mg/L)	0.63	0.58	0.10			0.10	
Iodide (μg/L)							
Potassium (mg/L)	362	297	60			4.2	
Magnesium (mg/L)	861	652.0	19.3			5.4	
Sodium (mg/L)	10156	9224	1545			124	
Silica (mg/L)	1.03	0.93	0.10			0.10	
Sulfate (mg/L)	2560	2201	32			1.1	
TDS (mg/L)	35963	31440	4195		59684	328	11562
Langelier Saturation Index	0.79	0.59	-2.04			-3.65	
Recovery (percent)	41		52			64	
Flux (gfd)			5.8			18.4	
Energy (kWh/KGal)						14.3	

> For column headings, the preceding digit denotes Stage 1 or 2, F: feed, P permeate, C: concentrate, R: return
> Pilot unit configuration: 4 vessels/5 elements in Stage 1 and 1 vessel/4 elements in Stage 2 (element area: 82 ft^2)
> Energy based on Eq. (2.1)

Table B.4
Pilot test results for FilmTec NF-90 at 580 psi in Stage 1, target salinity of 200 mg/L in Stage 2 permeate, and no Stage 1 permeate return.

5/11/2004	Water location						
Test 4	Raw	1F	1P/2F	1R	1C	2P	2C
Temperature (° F)	65.1	66	71.1		72.7	73.4	73.6
Pressure (psi)		581	170		573		145
Flow (gpm)	9.9	13.9	7.8		6.0	3.9	4.0
pH	8.01	8.02	7.69		7.89	7.99	7.41
Alkalinity (mg/L as $CaCO_3$)	117.3	84.9	7.9			3.1	
Boron (mg/L)	4.8	4.6	3.3			2.4	
Bromide (mg/L)	55.6	42.4	7.4				
Calcium (mg/L)	419	307.0	6.9			0.6	
Chloride (mg/L)	18939	14730	1838			106	
Fluoride (mg/L)	0.63	0.52	0.10			0.10	
Iodide (µg/L)							
Potassium (mg/L)	362	304	49			2.08	
Magnesium (mg/L)	861	633.0	15.2			4.1	
Sodium (mg/L)	10156	8244	1285			70	
Silica (mg/L)	1.03	0.84	0.10			0.10	
Sulfate (mg/L)	2560	1936	21			1.1	
TDS (mg/L)	35963	28071	3445		58714	180	7496
Langelier Saturation Index	0.83	0.58	-2.28			-3.30	
Recovery (percent)	39		56			49	
Flux (gfd)			6.6			15.7	
Energy (kWh/KGal)						16.9	

> For column headings, the preceding digit denotes Stage 1 or 2, F: feed, P permeate, C: concentrate, R: return
> Pilot unit configuration: 4 vessels/5 elements in Stage 1 and 1 vessel/4 elements in Stage 2 (element area: 82 ft^2)
> Energy based on Eq. (2.1)

Table B.5
Pilot test results for FilmTec NF-90 at 520 psi in Stage 1, target salinity of 350 mg/L in Stage 2 permeate, and 40 percent of Stage 1 permeate returned.

5/20/2004	Water location						
Test 21	Raw	1F	1P/2F	1R	1C	2P	2C
Temperature (° F)	65.7	64.9	65.7	66.6	67.5	69.4	69.8
Pressure (psi)	132	521	189		506		183
Flow (gpm)	13.9	17.5	3.9	3.3	10.4	3.5	0.3
pH	7.66	7.70	7.90	7.69	7.51	8.00	7.13
Alkalinity (mg/L as $CaCO_3$)	111.0	89.9				3.6	
Boron (mg/L)	4.8	4.4	2.5			2.2	
Bromide (mg/L)	59.3	41.8	4.9			1.1	
Calcium (mg/L)	385	315	5.33			1.18	
Chloride (mg/L)	19038	12723	1270				
Fluoride (mg/L)	0.60	0.53	0.04			0.10	
Iodide (µg/L)							
Potassium (mg/L)	387	297	27.9			5.80	
Magnesium (mg/L)	1217	994	15.2			3.54	
Sodium (mg/L)	9537	7578	668			158	
Silica (mg/L)	1.04	1.05	0.1			0.1	
Sulfate (mg/L)	2647	2096	32			3.7	
TDS (mg/L)	36163	30185	1938	3779	50955	349	14762
Langelier Saturation Index	0.42	0.29	-2.56			-3.00	
Recovery (percent)	26		22			92	
Flux (gfd)			3.4			15.8	
Energy (kWh/KGal)						18.1	

> For column headings, the preceding digit denotes Stage 1 or 2, F: feed, P permeate, C: concentrate, R: return
> Pilot unit configuration: 4 vessels/5 elements in Stage 1 and 1 vessel/4 elements in Stage 2 (element area: 82 ft^2)
> Energy based on Eq. (2.1)

Table B.6
Pilot test results for FilmTec NF-90 at 520 psi in Stage 1, target salinity of 200 mg/L in Stage 2 permeate, and 40 percent of Stage 1 permeate returned.

5/20/2004	Water location						
Test 22	Raw	1F	1P/2F	1R	1C	2P	2C
Temperature (° F)	67.3	62.96	64.4	65.7	67.5	69.3	70.0
Pressure (psi)	133	522	146		508		141
Flow (gpm)	13.3	17.1	4.2	3.2	10.0	3.3	0.6
pH	7.65	7.74	7.95	7.70	7.51	8.13	7.11
Alkalinity (mg/L as $CaCO_3$)	111.0	87.1				3.8	
Boron (mg/L)	4.8	4.3	2.4			2.2	
Bromide (mg/L)	59.3	42.6	4.2			0.6	
Calcium (mg/L)	385	305	4.89			1.34	
Chloride (mg/L)	19038	12675	1104			115	
Fluoride (mg/L)	0.6	0.51	0.1			0.1	
Iodide (μg/L)							
Potassium (mg/L)	387	285	25.8			6.06	
Magnesium (mg/L)	1217	962	13.8			3.51	
Sodium (mg/L)	9537	7563	621			77.5	
Silica (mg/L)	1.0	1.05	0.1			0.1	
Sulfate (mg/L)	2647	1984	26			1.1	
TDS (mg/L)	36163	29392	1831	3611	50732	194	8724
Langelier Saturation Index	0.43	0.28	-2.53			-2.77	
Recovery (percent)	26		25			85	
Flux (gfd)			3.7			15.6	
Energy (kWh/KGal)						18.6	

> For column headings, the preceding digit denotes Stage 1 or 2, F: feed, P permeate, C: concentrate, R: return
> Pilot unit configuration: 4 vessels/5 elements in Stage 1 and 1 vessel/4 elements in Stage 2 (element area: 82 ft^2)
> Energy based on Eq. (2.1)

Table B.7
Pilot test results for FilmTec NF-90 at 580 psi in Stage 1, target salinity of 200 mg/L in Stage 2 permeate, and 40 percent of Stage 1 permeate returned.

5/26/2004	Water location						
Test 23	Raw	1F	1P/2F	1R	1C	2P	2C
Temperature (° F)	66.2	65.66	66.0	66.6	68.0	68.5	68.9
Pressure (psi)	134	579	170		568		166
Flow (gpm)	12.3	16.4	4.9	3.3	8.3	4.0	0.8
pH	7.57	7.63	7.68	7.41	7.37	7.96	6.98
Alkalinity (mg/L as $CaCO_3$)	108.0	84.0	6.3		158.0	4.4	22.4
Boron (mg/L)	4.6	4.5	2.5	3.8	6.0	2.2	4.4
Bromide (mg/L)	64.3	46.9	21.6	8.31	81.8	0.54	17.7
Calcium (mg/L)	408	295	7.52	7.40	577	0.6	28.1
Chloride (mg/L)	19721	14882	1115	2547	29379	117	5472
Fluoride (mg/L)	0.56	0.51	0.1	0.30	0.58	0.1	0.1
Iodide (µg/L)							
Potassium (mg/L)	396	290	25.4	61.8	560	2.65	131
Magnesium (mg/L)	1291	927	21.2	20.1	1809	0.21	83.8
Sodium (mg/L)	9772	7891	622	1565	13477	72.1	3114
Silica (mg/L)	1.43	1.11	0.1	0.1	1.79	0.1	0.19
Sulfate (mg/L)	2792	2000	27.5	31.8	4202	1.1	142
TDS (mg/L)	36698	29458	1885	4201	55208	176	9473
Langelier Saturation Index	0.35	0.17	-2.54	-2.69	0.47	-3.23	-2.00
Recovery (percent)	33		30			83	
Flux (gfd)			4.3			17.9	
Energy (kWh/KGal)						17.7	

> For column headings, the preceding digit denotes Stage 1 or 2, F: feed, P permeate, C: concentrate, R: return
> Pilot unit configuration: 4 vessels/5 elements in Stage 1 and 1 vessel/4 elements in Stage 2 (element area: 82 ft^2)
> Energy based on Eq. (2.1)

Table B.8

Pilot test results for FilmTec NF-90 at 580 psi in Stage 1, target salinity of 350 mg/L in Stage 2 permeate, and 40 percent of Stage 1 permeate returned.

5/26/2004	Water location						
Test 24	Raw	1F	1P/2F	1R	1C	2P	2C
Temperature (° F)	62.6	63.14	64.9	64.0	66.7	68.4	69.6
Pressure (psi)	135	580	210		570		205
Flow (gpm)	12.5	16.0	4.6	3.2	8.3	4.2	0.3
pH	7.66	7.69	7.73	7.47	7.46	7.92	7.13
Alkalinity (mg/L as $CaCO_3$)	108.0	84.9	6.1			3.4	
Boron (mg/L)	4.6	4.6	2.5			2.3	
Bromide (mg/L)	64.3	50.5	15.7			0.66	
Calcium (mg/L)	408	294	9.23			0.6	
Chloride (mg/L)	19721	15732	1157			202	
Fluoride (mg/L)	0.6	0.45	0.1			0.1	
Iodide (µg/L)							
Potassium (mg/L)	396	295	25.5			2.70	
Magnesium (mg/L)	1291	922	26.5			2.70	
Sodium (mg/L)	9772	8092	613			145.2	
Silica (mg/L)	1.4	1.04	0.1			0.1	
Sulfate (mg/L)	2792	2141	27.4			1.1	
TDS (mg/L)	36698	29920	1869	4271	54611	331	15549
Langelier Saturation Index	0.40	0.20	-2.27			-3.40	
Recovery (percent)	34		29			93	
Flux (gfd)			4.0			18.8	
Energy (kWh/KGal)						16.8	

> For column headings, the preceding digit denotes Stage 1 or 2, F: feed, P permeate, C: concentrate, R: return

> Pilot unit configuration: 4 vessels/5 elements in Stage 1 and 1 vessel/4 elements in Stage 2 (element area: 82 ft^2)

> Energy based on Eq. (2.1)

Table B.9
Boron removal pilot test results for FilmTec NF-90 at 580 psi in Stage 1, pH 8.6 in Stage 2, 200 mg/L target salinity, and 20 percent of Stage 1 permeate returned.

6/28/2004	Water location						
Test 53	Raw	1F	1P/2F	1R	1C	2P	2C
Temperature (° F)	67.5	68.2	69.4	69.3	70.7	71.1	71.2
Pressure (psi)	132	580	174		569		162
Flow (gpm)	12.7	16.4	6.2	1.6	8.7	4.0	2.1
pH	8.01	8.08	8.21	8.03	7.81	8.58	7.42
Alkalinity (mg/L as $CaCO_3$)	122.1	98.5	8.6	12.8	176.4	4.1	13.8
Boron (mg/L)	4.7	4.5	2.9	4.1	5.8	2.3	4.0
Bromide (mg/L)	65.8	52.5	7.47	16.8	32.7	0.65	17.4
Calcium (mg/L)	425	323	7.67	12.6	602	0.6	20.5
Chloride (mg/L)	15993	9964	1510	3636	9658	118	4115
Fluoride (mg/L)	0.642	0.601	0.1	0.1	0.767	0.1	0.1
Iodide (µg/L)	41.9	34.9	8.0	19.2	64.8	0.25	17.0
Potassium (mg/L)	411	362	40.8	94.8	585	3.05	114
Magnesium (mg/L)	1339	1002	21.5	34.7	1873	0.25	58.1
Sodium (mg/L)	10947	9420	1025	2364	14561	76.7	2828
Silica (mg/L)	1.22	1.05	0.1	0.14		0.1	0.18
Sulfate (mg/L)	2675	1981	35.5	52.5	1112	1.1	70.2
TDS (mg/L)	37032	30779	3008	6400	54711	194	8453
Langelier Saturation Index	0.88	0.75	-1.76	-1.52	1.01	-2.68	-1.87
Recovery (percent)	32		38			66	
Flux (gfd)			5.4			17.8	
Energy (kWh/KGal)						17.6	

> For column headings, the preceding digit denotes Stage 1 or 2, F: feed, P permeate, C: concentrate, R: return
> Pilot unit configuration: 4 vessels/5 elements in Stage 1 and 1 vessel/4 elements in Stage 2 (element area: 82 ft^2)
> Energy based on Eq. (2.1)

Table B.10
Boron removal pilot test results for FilmTec NF-90 at 580 psi in Stage 1, pH 9.1 in Stage 2, 200 mg/L target salinity, and 20 percent of Stage 1 permeate returned.

6/28/2004	Water location						
Test 54	Raw	1F	1P/2F	1R	1C	2P	2C
Temperature (° F)	66.9	68.4	69.3	69.8	71.2	71.6	71.8
Pressure (psi)	132	581	170		570		160
Flow (gpm)	12.6	16.2	6.2	2.3	8.7	3.9	2.1
pH	8.07	8.11	8.30	8.05	7.84	9.05	7.92
Alkalinity (mg/L as $CaCO_3$)	121.7	99.7	8.6	11.4	178	5.7	15.4
Boron (mg/L)	4.6	4.5	2.9	4.1	5.8	2.4	4.1
Bromide (mg/L)	65.3	53.6	7.47	17.0	79.7	0.64	18.3
Calcium (mg/L)	422	330	7.42	12.9	604	0.6	20.0
Chloride (mg/L)	17055	8997	1693	3730	22161	117	4226
Fluoride (mg/L)	0.601	0.567	0.1	0.1	0.64	0.1	0.1
Iodide (μg/L)	44.3	42.3	9.9	21.7	67.0	0.47	19.9
Potassium (mg/L)	422	364	41.3	98.8	580	3.03	109
Magnesium (mg/L)	1329	1107	20.8	35.6	1880	0.26	559
Sodium (mg/L)	11188	9442	1018	2404	14967	77.2	2668
Silica (mg/L)	1.47	1.08	0.1	0.12	1.61	0.1	0.16
Sulfate (mg/L)	2655	2004	33.2	53.2	2796	1.1	70.8
TDS (mg/L)	36965	31109	3067	6591	54924	194	8160
Langelier Saturation Index	0.93	0.80	-1.68	-1.53	1.05	-2.06	-1.33
Recovery (percent)	32		38			65	
Flux (gfd)			5.4			17.7	
Energy (kWh/KGal)						17.8	

> For column headings, the preceding digit denotes Stage 1 or 2, F: feed, P permeate, C: concentrate, R: return
> Pilot unit configuration: 4 vessels/5 elements in Stage 1 and 1 vessel/4 elements in Stage 2 (element area: 82 ft^2)
> Energy based on Eq. (2.1)

Table B.11
Boron removal pilot test results for FilmTec NF-90 at 580 psi in Stage 1, pH 9.3 in Stage 2, 200 mg/L target salinity, and 20 percent of Stage 1 permeate returned.

6/28/2004	Water location						
Test 55	Raw	1F	1P/2F	1R	1C	2P	2C
Temperature (° F)	66.0	66.9	68.4	68.9	70.3	70.2	70.9
Pressure (psi)	134	581	180		571		168
Flow (gpm)	12.8	16.4	6.2	1.5	8.7	4.1	2.1
pH	8.09	8.20	8.32	8.14	7.94	9.33	8.96
Alkalinity (mg/L as $CaCO_3$)	122.8	102.1	8.6	12.6	179.2	6.3	26
Boron (mg/L)	4.6	4.6	2.9	4.1	5.9	1.8	5.0
Bromide (mg/L)	65.0	52.0	7.49	15.7	72.0	0.57	18.4
Calcium (mg/L)	437	328	7.20	12.7	639	0.6	19.7
Chloride (mg/L)	15579	12879	1566	3501	20393	105	4254
Fluoride (mg/L)	0.685	0.644	0.1	0.1	0.693	0.1	0.1
Iodide (μg/L)	58.2	46.0	8.3	15.3	103.8	0.73	29.9
Potassium (mg/L)	421	349	41.4	95.1	611	2.60	106
Magnesium (mg/L)	1370	1024	20.3	34.8	1984	0.26	55.6
Sodium (mg/L)	11201	8990	1008	2294	15542	68.0	2585
Silica (mg/L)	1.10	0.88	0.1	0.12	1.84	0.1	0.17
Sulfate (mg/L)	2632	1963	33.1	52.8	2554	1.1	72.3
TDS (mg/L)	37099	31109	2971	6591	54711	170	8424
Langelier Saturation Index	0.95	0.88	-1.69	-1.41	1.16	-1.77	-0.08
Recovery (percent)	32		38			66	
Flux (gfd)			5.4			18.0	
Energy (kWh/KGal)						17.3	

> For column headings, the preceding digit denotes Stage 1 or 2, F: feed, P permeate, C: concentrate, R: return
> Pilot unit configuration: 4 vessels/5 elements in Stage 1 and 1 vessel/4 elements in Stage 2 (element area: 82 ft^2)
> Energy based on Eq. (2.1)

Table B.12
Boron removal pilot test results for FilmTec NF-90 at 580 psi in Stage 1, pH 9.7 in Stage 2, 200 mg/L target salinity, and 20 percent of Stage 1 permeate returned.

6/28/2004	Water location						
Test 56	Raw	1F	1P/2F	1R	1C	2P	2C
Temperature (° F)	66.4	67.5	69.3	68.9	70.7	71.4	71.6
Pressure (psi)	133	580	174		568		162
Flow (gpm)	13.2	16.8	6.1	1.5	9.2	4.0	2.1
pH	8.20	8.33	8.49	8.30	8.09	9.65	9.47
Alkalinity (mg/L as $CaCO_3$)	124.8	105.2	8.5	12.3	182.8	11.8	35.5
Boron (mg/L)	4.7	4.7	2.8	4.1	5.7	1.3	5.2
Bromide (mg/L)	66.7	55.5	7.03	14.2	59.5	0.54	19.6
Calcium (mg/L)	422	339	6.80	11.2	599	0.6	19.3
Chloride (mg/L)	15075	11285	1423	3128	16960	98.6	4218
Fluoride (mg/L)	0.584	0.546	0.1	0.1	0.571.	0.1	0.1
Iodide (μg/L)	48.3	48.2	9.5	15.0	103.8	0.25	21.3
Potassium (mg/L)	422	364	39.2	86.9	571	2.47	104
Magnesium (mg/L)	1320	1057	19.2	30.9	1862	0.26	56.3
Sodium (mg/L)	10747	9242	934	2083	14699	64.5	1494
Silica (mg/L)	1.27	1.06	0.1	0.13	1.68	0.1	0.18
Sulfate (mg/L)	2718	2093	30.8	46.4	2057	1.1	72.0
TDS (mg/L)	37032	31109	2838	5990	54142	160	8107
Langelier Saturation Index	1.06	1.04	-1.54	-1.32	1.30	-1.14	0.57
Recovery (percent)	30		36			66	
Flux (gfd)			5.4			17.6	
Energy (kWh/KGal)						17.7	

> For column headings, the preceding digit denotes Stage 1 or 2, F: feed, P permeate, C: concentrate, R: return
> Pilot unit configuration: 4 vessels/5 elements in Stage 1 and 1 vessel/4 elements in Stage 2 (element area: 82 ft^2)
> Energy based on Eq. (2.1)

Table B.13
Boron removal pilot test results for FilmTec NF-90 at 580 psi in Stage 1, pH 10.3 in Stage 2, 200 mg/L target salinity, and 20 percent of Stage 1 permeate returned.

6/28/2004	Water location						
Test 57	Raw	1F	1P/2F	1R	1C	2P	2C
Temperature (° F)	67.1	67.5	68.5	68.9	71.1	71.8	72.5
Pressure (psi)	132	579	162		565		154
Flow (gpm)	12.9	16.5	6.2	1.4	9.0	3.9	2.2
pH	8.25	8.44	8.64	8.43	8.17	10.27	9.99
Alkalinity (mg/L as $CaCO_3$)	125.6	107.3	8.0	11.9	186.5	9.2	46.9
Boron (mg/L)	4.6	4.8	2.8	4.0	6.5	0.9	6.3
Bromide (mg/L)	61.9	53.4	6.87	14.2	84.2	0.53	16.3
Calcium (mg/L)	423	334	6.69	11.0	595	0.6	18.2
Chloride (mg/L)	15850	10198	1401	3071	22234	96.2	3757
Fluoride (mg/L)	0.573	0.535	0.1	0.1	0.624	0.1	0.1
Iodide (µg/L)	47.7	49.0	8.0	15.7	105.0	0.25	20.3
Potassium (mg/L)	425	358	38.7	84.3	568	2.30	99
Magnesium (mg/L)	1319	1042	18.9	30.4	1851	0.26	52.8
Sodium (mg/L)	10835	9199	929	2026	14578	63.8	2372
Silica (mg/L)	2.25	1.11	0.1	0.1	1.73	0.1	0.14
Sulfate (mg/L)	2499	2031	30.7	44.9	2780	1.1	61.8
TDS (mg/L)	36631	30383	2742	5553	53503	160	7583
Langelier Saturation Index	1.12	1.15	-1.43	-1.21	1.39	-0.63	1.20
Recovery (percent)	31		38			64	
Flux (gfd)			5.4			17.4	
Energy (kWh/KGal)						17.8	

> For column headings, the preceding digit denotes Stage 1 or 2, F: feed, P permeate, C: concentrate, R: return
> Pilot unit configuration: 4 vessels/5 elements in Stage 1 and 1 vessel/4 elements in Stage 2 (element area: 82 ft^2)
> Energy based on Eq. (2.1)

Table B.14
Boron removal pilot test results for FilmTec NF-90 at 580 psi in Stage 1, pH 10.7 in Stage 2, 200 mg/L target salinity, and 20 percent of Stage 1 permeate returned.

6/28/2004	Water location						
Test 58	Raw	1F	1P/2F	1R	1C	2P	2C
Temperature (° F)	67.6	69.1	70.2	70.0	72.1	73.2	73.6
Pressure (psi)	133	579	164		567		152
Flow (gpm)	13.0	16.6	6.2	1.4	9.1	3.9	2.2
pH	8.33	8.51	8.60	8.45	8.26	10.72	10.28
Alkalinity (mg/L as $CaCO_3$)	130.1	112	8.7	11.9	199	15.3	59.3
Boron (mg/L)	4.7	4.8	2.8	4.0	6.5	0.7	6.7
Bromide (mg/L)	63.4	53.4	7.29	16.0	61.3	0.58	18.5
Calcium (mg/L)	437	355	6.83	11.1	605	0.6	18.1
Chloride (mg/L)	14886	9174	1517	3436	17887	107	3967
Fluoride (mg/L)	0.531	0.496	0.1	0.1	0.566	0.1	0.1
Iodide (μg/L)	48.6	46.6	8.1	19.0	104.6	0.25	27.8
Potassium (mg/L)	428	356	40.5	96.5	578	2.78	104
Magnesium (mg/L)	1357	1106	19.2	29.7	1880	0.31	52.6
Sodium (mg/L)	10934	9131	975	2341	15346	74.2	2522
Silica (mg/L)	1.07	1.17	0.1	0.14	1.74	0.1	0.18
Sulfate (mg/L)	2718	2033	30.8	41.2	2173	1.1	62.2
TDS (mg/L)	36898	30779	2891	6075	53503	195	7992
Langelier Saturation Index	1.24	1.28	-1.41	-1.17	1.52	0.05	1.59
Recovery (percent)	30		37			64	
Flux (gfd)			5.4			17.4	
Energy (kWh/KGal)						17.8	

> For column headings, the preceding digit denotes Stage 1 or 2, F: feed, P permeate, C: concentrate, R: return
> Pilot unit configuration: 4 vessels/5 elements in Stage 1 and 1 vessel/4 elements in Stage 2 (element area: 82 ft^2)
> Energy based on Eq. (2.1)

Table B.15
Pilot test results for Saehan NE-90 at 520 psi in Stage 1, target salinity of 350 mg/L in Stage 2 permeate, and no Stage 1 permeate return.

7/29/2004	Water location						
Test 72	Raw	1F	1P/2F	1R	1C	2P	2C
Temperature (^{o} F)	66.9	69.8	71.6		73.6	73.4	73.9
Pressure (psi)	130	520	199		509		130
Flow (gpm)	10.3	18.1	12.0		6.0	4.3	7.8
pH	7.73	7.84	8.11		7.53	8.82	7.62
Alkalinity (mg/L as $CaCO_3$)	111	69.0	6.38		184	6.98	9.55
Boron (mg/L)	4.7	4.5	3.8		5.8	3.3	4.1
Bromide (mg/L)	60.8	43.7	12.8		82.9	0.91	21.1
Calcium (mg/L)	406	239	16.5		281	0.42	24.0
Chloride (mg/L)	19614	12004	3600		23899	187	5058
Fluoride (mg/L)	0.596	0.467	0.1		0.689	0.1	0.2
Iodide (µg/L)	56	56	11		89	0.686	17
Potassium (mg/L)	402	264	78.0		260	4.28	114
Magnesium (mg/L)	1221	699	35.1		832	0.43	49.2
Sodium (mg/L)	10700	6708	2064		6455	117	3134
Silica (mg/L)	0.86	0.5	0.1		1.212	0.1	0.13
Sulfate (mg/L)	2461	1463	28.1		2921	1.1	35.1
TDS (mg/L)	34759	24702	5644		58405	338	9150
Langelier Saturation Index	0.53	0.25	-1.59		0.44	-2.30	-1.74
Recovery (percent)	42		66			36	
Flux (gfd)			10.5			18.7	
Energy (kWh/KGal)						19.7	

> For column headings, the preceding digit denotes Stage 1 or 2, F: feed, P permeate, C: concentrate, R: return
> Pilot unit configuration: 4 vessels/5 elements in Stage 1 and 1 vessel/4 elements in Stage 2 (element area: 82 ft^2)
> Energy based on Eq. (2.1)

Table B.16
Pilot test results for Saehan NE-90 at 520 psi in Stage 1, target salinity of 200 mg/L in Stage 2 permeate, and 40 percent of Stage 1 permeate returned.

7/29/2004	Water location						
Test 75	Raw	1F	1P/2F	1R	1C	2P	2C
Temperature (° F)	68.9	69.3	70.9	71.2	73.4	73.8	74.3
Pressure (psi)	130	520	142		509		122
Flow (gpm)	10.3	17.7	7.0	4.4	6.3	4.0	3.0
pH	7.71	7.75	8.29	7.84	7.46	8.90	7.65
Alkalinity (mg/L as $CaCO_3$)	111	69.6	7.8			5.2	
Boron (mg/L)	4.7	4.6	3.4			3.4	
Bromide (mg/L)	60.8	47.0	5.96			0.48	
Calcium (mg/L)	406	243	6.11			0.92	
Chloride (mg/L)	19614	13176	1606			102	
Fluoride (mg/L)	0.6	0.463	0.1			0.1	
Iodide (µg/L)							
Potassium (mg/L)	402	280	39.0			2.42	
Magnesium (mg/L)	1221	708	17.1			0.43	
Sodium (mg/L)	10700	7221	984			64.6	
Silica (mg/L)	0.86	0.59	0.1			0.1	
Sulfate (mg/L)	2461	1536	14.4			1.13	
TDS (mg/L)	34759	25825	3051			176	7490
Langelier Saturation Index	0.53	0.16	-1.74			-1.98	
Recovery (percent)	39		40			57	
Flux (gfd)			6.1			17.6	
Energy (kWh/KGal)						19.1	

> For column headings, the preceding digit denotes Stage 1 or 2, F: feed, P permeate, C: concentrate, R: return
> Pilot unit configuration: 4 vessels/5 elements in Stage 1 and 1 vessel/4 elements in Stage 2 (element area: 82 ft^2)
> Energy based on Eq. (2.1)

Table B.17
Pilot test results for Saehan NE-90 at 520 psi in Stage 1, target salinity of 350 mg/L in Stage 2 permeate, and 40 percent of Stage 1 permeate returned.

7/29/2004	Water location						
Test 76	Raw	1F	1P/2F	1R	1C	2P	2C
Temperature (° F)	69.8	69.3	70.2	70.9	72.5	72.9	73.4
Pressure (psi)	130	520	198		508		180
Flow (gpm)	12.4	18.1	6.2	4.4	7.5	4.9	1.3
pH	7.66	7.74	8.32	7.87	7.46	8.74	7.51
Alkalinity (mg/L as $CaCO_3$)	111	82.0	6.1			4.7	
Boron (mg/L)	4.7	4.7	3.5			3.0	
Bromide (mg/L)	60.8	52.2	7.95			1.01	
Calcium (mg/L)	406	279	7.74			0.88	
Chloride (mg/L)	19614	14486	1880			211	
Fluoride (mg/L)	0.6	0.499	0.1			0.1	
Iodide (µg/L)							
Potassium (mg/L)	402	228	45.6			4.76	
Magnesium (mg/L)	1221	821	20.5			0.66	
Sodium (mg/L)	10700	5776	1161			132	
Silica (mg/L)	0.86	0.59	0.1			0.1	
Sulfate (mg/L)	2461	1779	18.0			1.14	
TDS (mg/L)	34759	28996	3493			355	
Langelier Saturation Index	0.49	0.28	-1.72			-2.23	
Recovery (percent)	40		34			79	
Flux (gfd)			5.4			21.5	
Energy (kWh/KGal)						16.0	

> For column headings, the preceding digit denotes Stage 1 or 2, F: feed, P permeate, C: concentrate, R: return
> Pilot unit configuration: 4 vessels/5 elements in Stage 1 and 1 vessel/4 elements in Stage 2 (element area: 82 ft^2)
> Energy based on Eq. (2.1)

Table B.18
Pilot test results for Saehan NE-90 at 470 psi in Stage 1, target salinity of 350 mg/L in Stage 2 permeate, and no Stage 1 permeate return.

8/4/2004	Water location						
Test 79	Raw	1F	1P/2F	1R	1C	2P	2C
Temperature (° F)	67.8	70.0	71.6		72.9	73.2	73.6
Pressure (psi)	130	470	170		453		111
Flow (gpm)	10.7	18.0	10.8		7.2	3.5	7.3
pH	7.66	7.76	8.13		7.63	8.86	7.74
Alkalinity (mg/L as $CaCO_3$)	64.2	69.7	11.3			6.1	
Boron (mg/L)	4.7	4.2	3.6			3.3	
Bromide (mg/L)	59.2	40.7	15.0			0.95	
Calcium (mg/L)	402	237	16.0			0.6	
Chloride (mg/L)	19572	12506	3623			199	
Fluoride (mg/L)	0.5	0.52	0.1			0.1	
Iodide (μg/L)							
Potassium (mg/L)	352	270	109			4.64	
Magnesium (mg/L)	1063	678	37.2			0.46	
Sodium (mg/L)	10600	7237	2092			125	
Silica (mg/L)	0.64	0.77	0.1			0.1	
Sulfate (mg/L)	2343	1432	29			1.1	
TDS (mg/L)	35980	25429	5601			364	
Langelier Saturation Index	0.25	0.17	-1.34			-2.16	
Recovery (percent)	33		60			32	
Flux (gfd)			9.5			15.4	
Energy (kWh/KGal)						20.3	

> For column headings, the preceding digit denotes Stage 1 or 2, F: feed, P permeate, C: concentrate, R: return
> Pilot unit configuration: 4 vessels/5 elements in Stage 1 and 1 vessel/4 elements in Stage 2 (element area: 82 ft^2)
> Energy based on Eq. (2.1)

Table B.19
Pilot test results for Saehan NE-90 at 470 psi in Stage 1, target salinity of 200 mg/L in Stage 2 permeate, and 40 percent of Stage 1 permeate returned.

8/4/2004	Water location						
Test 80	Raw	1F	1P/2F	1R	1C	2P	2C
Temperature (° F)	66.6	68.5	70.2	70.9	72.3	72.3	73.0
Pressure (psi)	130	470	125		455		100
Flow (gpm)	10.2	17.8	6.8	4.1	6.8	3.4	3.5
pH	7.71	7.84	8.32	7.87	7.52	8.78	7.67
Alkalinity (mg/L as $CaCO_3$)	64.2	68.1	6.7			5.3	
Boron (mg/L)	4.7	4.3	3.2			3.0	
Bromide (mg/L)	59.2	45.3	6.57			0.49	
Calcium (mg/L)	402	245	6.50			0.6	
Chloride (mg/L)	19572	12784	1696			99.7	
Fluoride (mg/L)	0.5	0.49	0.1			0.1	
Iodide (μg/L)							
Potassium (mg/L)	325	238	42.9			2.36	
Magnesium (mg/L)	1063	709	18.2			0.32	
Sodium (mg/L)	10600	6033	1059			63.3	
Silica (mg/L)	0.64	0.63	0.1			0.1	
Sulfate (mg/L)	2343	1536	15.6			1.1	
TDS (mg/L)	35980	26631	3108			184	
Langelier Saturation Index	0.28	0.24	-1.75			-2.27	
Recovery (percent)	33		38			49	
Flux (gfd)			6.0			14.7	
Energy (kWh/KGal)						19.8	

> For column headings, the preceding digit denotes Stage 1 or 2, F: feed, P permeate, C: concentrate, R: return
> Pilot unit configuration: 4 vessels/5 elements in Stage 1 and 1 vessel/4 elements in Stage 2 (element area: 82 ft^2)
> Energy based on Eq. (2.1)

Table B.20
Pilot test results for Saehan NE-90 at 470 psi in Stage 1, target salinity of 350 mg/L in Stage 2 permeate, and 40 percent of Stage 1 permeate returned.

8/4/2004	Water location						
Test 81	Raw	1F	1P/2F	1R	1C	2P	2C
Temperature (° F)	67.6	68.9	70.7	71.2	72.9	73.8	74.5
Pressure (psi)	130	470	160		453		150
Flow (gpm)	11.7	17.5	5.7	4.0	7.8	3.9	1.8
pH	7.73	7.72	8.27	7.88	7.53	8.67	7.49
Alkalinity (mg/L as $CaCO_3$)	64.2	79.1	9.2			4.5	
Boron (mg/L)	4.7	4.5	3.3			3.1	
Bromide (mg/L)	59.2	95.9	8.57			0.99	
Calcium (mg/L)	402	286	8.62			0.63	
Chloride (mg/L)	19572	26218	2077			203	
Fluoride (mg/L)	0.5	0.53	0.1			0.1	
Iodide (µg/L)							
Potassium (mg/L)	352	167	49.2			4.68	
Magnesium (mg/L)	1063	830	23.4			0.56	
Sodium (mg/L)	10600	4051	1244			128	
Silica (mg/L)	0.64	0.91	0.1			0.1	
Sulfate (mg/L)	2343	2926	19.4			1.1	
TDS (mg/L)	35602	30237	3813			363	
Langelier Saturation Index	0.32	0.25	-1.55			-2.46	
Recovery (percent)	33		33			68	
Flux (gfd)			5.0			17.1	
Energy (kWh/KGal)						16.7	

> For column headings, the preceding digit denotes Stage 1 or 2, F: feed, P permeate, C: concentrate, R: return
> Pilot unit configuration: 4 vessels/5 elements in Stage 1 and 1 vessel/4 elements in Stage 2 (element area: 82 ft^2)
> Energy based on Eq. (2.1)

Table B.21
Boron removal pilot test results for Saehan NE-90 at 520 psi in Stage 1, pH 8.9 in Stage 2, 250 mg/L target salinity, and 16 percent of Stage 1 permeate returned.

8/25/2004	Water location						
Test 95	Raw	1F	1P/2F	1R	1C	2P	2C
Temperature (° F)	62.8	64.4	66.2	66.2	67.8	68.2	68.7
Pressure (psi)	130	521	191		505		165
Flow (gpm)	1.27	17.7	8.0	1.6	8.2	4.5	3.4
pH	7.78	7.86	8.37	7.96	7.59	8.89	7.70
Alkalinity (mg/L as $CaCO_3$)	109	83.2	7.2	18.2	168	2.9	10.9
Boron (mg/L)	4.7	4.7	3.4	4.9	6.3	2.8	4.3
Bromide (mg/L)	68.4	52.3	10.4	32.3	100	0.7	23.3
Calcium (mg/L)	430	321	10.9	32.9	551	0.6	25.3
Chloride (mg/L)	18437	14141	2530	7520	27504	128	5659
Fluoride (mg/L)							
Iodide (µg/L)	59.8	49.2	8.8	23.0	88.5	0.5	18.4
Potassium (mg/L)	416	329	50.4	154	592	3.0	124
Magnesium (mg/L)	1334	955	29.1	91.7	1982	0.4	62.4
Sodium (mg/L)	10466	8727	1361	4026	15337	87.4	3195
Silica (mg/L)	0.84	0.71	0.1	0.1	1.18	0.1	0.2
Sulfate (mg/L)	2872	2018	30.0	72.6	3481	1.1	54.8
TDS (mg/L)	36230	29590	4898	11021	56203	242	9720
Langelier Saturation Index	0.55	0.41	-1.51	-1.08	0.69	-2.49	-1.64
Recovery (percent)	36		45			57	
Flux (gfd)			7.0			20.0	
Energy (kWh/KGal)						16.5	

> For column headings, the preceding digit denotes Stage 1 or 2, F: feed, P permeate, C: concentrate, R: return
> Pilot unit configuration: 4 vessels/5 elements in Stage 1 and 1 vessel/4 elements in Stage 2 (element area: 82 ft^2)
> Energy based on Eq. (2.1)

Table B.22
Boron removal pilot test results for Saehan NE-90 at 520 psi in Stage 1, pH 9.3 in Stage 2, 250 mg/L target salinity, and 16 percent of Stage 1 permeate returned.

8/24/2004	Water location						
Test 96	Raw	1F	1P/2F	1R	1C	2P	2C
Temperature (° F)	61.2	63.0	66.4	66.0	67.8	68.2	68.5
Pressure (psi)	130	521	190		506		165
Flow (gpm)	12.8	17.7	8.0	1.5	8.2	4.6	3.4
pH	7.78	8.04	8.51	8.08	7.73	9.30	7.96
Alkalinity (mg/L as $CaCO_3$)	116	84.2	8.5	17.4	169	6.8	12.8
Boron (mg/L)	4.8	4.7	3.4	4.8	6.0	2.8	4.2
Bromide (mg/L)	56.2	49.2	8.5	31.3	204	0.6	21.8
Calcium (mg/L)	433	329	10.3	34.2	563	1.0	23.9
Chloride (mg/L)	15132	13129	2212	7270	55562	116	5522
Fluoride (mg/L)							
Iodide (μg/L)	59.7	49.1	8.3	21.5	88.5	0.4	17.9
Potassium (mg/L)	397	333	46.1	150	607	2.8	119
Magnesium (mg/L)	1316	967	27.4	84.8	2059	0.6	59.8
Sodium (mg/L)	10241	8440	1223	3952	15957	81.9	3010
Silica (mg/L)	0.83	0.62	0.1	0.1	1.28	0.1	0.2
Sulfate (mg/L)	2856	2045	27.0	76.7	3815	1.1	52.3
TDS (mg/L)	36096	29326	4169	10297	56132	225	9421
Langelier Saturation Index	0.56	0.59	-1.31	-0.96	0.84	-1.48	-1.33
Recovery (percent)	36		45			58	
Flux (gfd)			7.0			20.2	
Energy (kWh/KGal)						16.2	

> For column headings, the preceding digit denotes Stage 1 or 2, F: feed, P permeate, C: concentrate, R: return
> Pilot unit configuration: 4 vessels/5 elements in Stage 1 and 1 vessel/4 elements in Stage 2 (element area: 82 ft^2)
> Energy based on Eq. (2.1)

Table B.23

Boron removal pilot test results for Saehan NE-90 at 520 psi in Stage 1, pH 9.5 in Stage 2, 250 mg/L target salinity, and 16 percent of Stage 1 permeate returned.

8/24/2004	Water location						
Test 97	Raw	1F	1P/2F	1R	1C	2P	2C
Temperature (° F)	63.5	64.4	66.2	66.6	68.2	68.4	68.9
Pressure (psi)	130	520	200		501		170
Flow (gpm)	12.8	17.7	8.0	1.5	8.2	4.6	3.4
pH	8.01	8.09	8.54	8.20	7.80	9.53	8.59
Alkalinity (mg/L as $CaCO_3$)	110	85.1	9.1	14.2	170	11.3	17.1
Boron (mg/L)	4.8	4.8	3.6	4.9	5.7	2.7	4.7
Bromide (mg/L)	68.2	62.2	10.0	33.3	89.0	0.6	21.2
Calcium (mg/L)	430	334	11.2	35.7	651	0.6	25.8
Chloride (mg/L)	18564	16364	2556	7712	23733	132	5915
Fluoride (mg/L)							
Iodide (µg/L)	59.0	49.4	9.0	22.2	88.5	0.5	19.2
Potassium (mg/L)	409	338	53.6	155	664	3.1	130
Magnesium (mg/L)	1337	973	29.5	89.2	2280	0.5	62.5
Sodium (mg/L)	10348	8511	1427	4142	17651	92.4	3338
Silica (mg/L)	0.82	0.80	0.1	0.24	1.28	0.1	0.19
Sulfate (mg/L)	2744	2042	28.4	78.5	3978	1.1	46.3
TDS (mg/L)	38252	30029	4562	10957	59759	247	10739
Langelier Saturation Index	0.56	0.59	-1.31	-0.96	0.98	-1.48	-0.54
Recovery (percent)	36		45			58	
Flux (gfd)			7.0			20.2	
Energy (kWh/KGal)						16.3	

> For column headings, the preceding digit denotes Stage 1 or 2, F: feed, P permeate, C: concentrate, R: return

> Pilot unit configuration: 4 vessels/5 elements in Stage 1 and 1 vessel/4 elements in Stage 2 (element area: 82 ft^2)

> Energy based on Eq. (2.1)

Table B.24
Boron removal pilot test results for Saehan NE-90 at 520 psi in Stage 1, pH 9.7 in Stage 2, 250 mg/L target salinity, and 16 percent of Stage 1 permeate returned.

8/24/2004	Water location						
Test 98	Raw	1F	1P/2F	1R	1C	2P	2C
Temperature (° F)	61.7	62.8	66.4	65.1	67.1	68.2	68.4
Pressure (psi)	130	520	190		502		170
Flow (gpm)	12.8	17.7	8.0	1.5	8.2	4.6	3.4
pH	7.92	8.13	8.62	8.23	7.81	9.74	9.25
Alkalinity (mg/L as $CaCO_3$)	111	88.7	10.5	17.6	178	7.5	43.8
Boron (mg/L)	4.7	5.0	3.5	5.0	6.9	1.5	6.0
Bromide (mg/L)	71.6	63.6	9.5	30.2	120	0.5	17.9
Calcium (mg/L)	395	325	10.5	33.7	584	0.6	24.3
Chloride (mg/L)	19234	16955	2456	7030	32350	109	5476
Fluoride (mg/L)							
Iodide (μg/L)	59.0	48.9	8.3	20.3	87.2	0.4	17.9
Potassium (mg/L)	414	330	51.8	143	592	2.7	118
Magnesium (mg/L)	1320	943	27.7	84.3	2005	0.4	60.6
Sodium (mg/L)	10355	8059	1375	3804	15571	79.1	3008
Silica (mg/L)	0.83	0.69	0.1	0.19	1.19	0.1	0.2
Sulfate (mg/L)	2807	2089	29.4	82.1	4140	1.1	44.6
TDS (mg/L)	38322	29683	4221	10946	59311	224	9983
Langelier Saturation Index	0.65	0.70	-1.10	-0.83	0.96	-1.22	0.50
Recovery (percent)	36		45			58	
Flux (gfd)			7.0			20.2	
Energy (kWh/KGal)						16.1	

> For column headings, the preceding digit denotes Stage 1 or 2, F: feed, P permeate, C: concentrate, R: return
> Pilot unit configuration: 4 vessels/5 elements in Stage 1 and 1 vessel/4 elements in Stage 2 (element area: 82 ft^2)
> Energy based on Eq. (2.1)

Table B.25
Boron removal pilot test results for Saehan NE-90 at 520 psi in Stage 1, pH 10.0 in Stage 2, 250 mg/L target salinity, and 16 percent of Stage 1 permeate returned.

8/24/2004	Water location						
Test 99	Raw	1F	1P/2F	1R	1C	2P	2C
Temperature (° F)	65.5	65.3	66.9	66.0	68.5	69.3	69.8
Pressure (psi)	130	520	190		502		165
Flow (gpm)	12.8	17.7	8.0	1.5	8.2	4.6	3.4
pH	8.01	8.24	8.69	8.36	7.92	9.97	9.56
Alkalinity (mg/L as $CaCO_3$)	113	91.4	5.4	16.1	186	8.6	44.2
Boron (mg/L)	5.1	5.3	3.5	4.9	6.7	1.2	6.5
Bromide (mg/L)	71.4	55.2	9.7	29.3	224	0.6	19.4
Calcium (mg/L)	434	312	18.8	33.5	615	1.0	24.6
Chloride (mg/L)	19693	15501	2506	6899	60861	118	5499
Fluoride (mg/L)							
Iodide (μg/L)	59.8	48.5	8.4	22.4	86.4	0.5	18.4
Potassium (mg/L)	377	311	52.4	138	625	2.9	121
Magnesium (mg/L)	1379	875	28.0	78.6	2131	0.4	60.3
Sodium (mg/L)	9549	8510	1388	3644	16388	85.1	3113
Silica (mg/L)	0.88	0.70	0.1	0.22	1.21	0.1	0.22
Sulfate (mg/L)	2781	2115	30.4	77.8	4238	1.1	45.7
TDS (mg/L)	38392	31139	4333	11063	59460	237	10152
Langelier Saturation Index	0.83	0.83	-1.06	-0.72	1.12	-0.72	0.83
Recovery (percent)	36		45			58	
Flux (gfd)			7.0			20.2	
Energy (kWh/KGal)						16.1	

> For column headings, the preceding digit denotes Stage 1 or 2, F: feed, P permeate, C: concentrate, R: return
> Pilot unit configuration: 4 vessels/5 elements in Stage 1 and 1 vessel/4 elements in Stage 2 (element area: 82 ft^2)
> Energy based on Eq. (2.1)

Table B.26
Boron removal pilot test results for Saehan NE-90 at 520 psi in Stage 1, pH 10.8 in Stage 2, 250 mg/L target salinity, and 16 percent of Stage 1 permeate returned.

8/24/2004	Water location						
Test 100	Raw	1F	1P/2F	1R	1C	2P	2C
Temperature (° F)	64.2	64.2	66.0	66.0	67.8	68.2	68.9
Pressure (psi)	132	520	190		502		162
Flow (gpm)	12.9	18.0	8.0	1.7	8.2	4.7	3.4
pH	8.32	8.59	8.90	8.60	8.26	10.83	10.23
Alkalinity (mg/L as $CaCO_3$)	117	95.2	8.5	15.4	191	16.3	56.6
Boron (mg/L)	4.6	5.1	3.2	4.9	7.2	0.6	6.8
Bromide (mg/L)	69.8	55.6	9.3	30.6	132	0.5	19.7
Calcium (mg/L)	413	324	10.5	32.6	508	0.6	23.4
Chloride (mg/L)	18424	14746	2307	7254	35454	101	5327
Fluoride (mg/L)							
Iodide (μg/L)	59.6	48.4	8.0	19.9	87.5	0.4	17.4
Potassium (mg/L)	401	320	47.9	142	566	2.6	113
Magnesium (mg/L)	1348	928	27.3	80.9	1945	0.5	59.9
Sodium (mg/L)	10335	8479	1248	3786	14829	78.0	3024
Silica (mg/L)	0.83	0.64	0.1	0.22	1.1	0.1	0.21
Sulfate (mg/L)	2910	1639	31.7	79.7	4637	1.1	48.3
TDS (mg/L)	38533	30654	4036	10398	60132	230	9626
Langelier Saturation Index	1.12	1.21	-0.92	-0.51	1.38	0.20	1.58
Recovery (percent)	36		44			58	
Flux (gfd)			7.0			20.4	
Energy (kWh/KGal)						16.1	

> For column headings, the preceding digit denotes Stage 1 or 2, F: feed, P permeate, C: concentrate, R: return
> Pilot unit configuration: 4 vessels/5 elements in Stage 1 and 1 vessel/4 elements in Stage 2 (element area: 82 ft^2)
> Energy based on Eq. (2.1)

REFERENCES

Abdel-Jawad M., E.E.F. El-Sayed, S. Ebrahim, A. Al-Saffar, M. Safar, M. Tabtabaei, and G. Al-Nuwaibit. 2001. Fifteen Years of R&D Program in Seawater Desalination at KISR: Part II. RO System Performance. *Desalination* 135:135-167

Adham S.S., R.S. Trussell, P.F. Gagliardo, and R.R. Trussell. 1998. Rejection of MS-2 Virus by RO Membranes. *Journal AWWA* 90(9):130-135

Akbari A., S. Desclaux, J.C. Remigy and P. Aptel. 2002. Treatment of Textile Dye Effluents Using a New Photografted Nanofiltration Membrane. *Desalination* 149:101-107

Allegeir S.C. and R.S. Summers 1995. Evaluation NF for DBP Control with the RBSMT. *Journal AWWA* 87(3):87-99.

Andrews-Tate, Cynthia <Cynthia_Andrews-Tate@lbwater.org >. 2004. TOC content of LBWD water. Email to Catherine Harrison <cheine@unr.nevada.edu>

AWWA. 1999. *Water Quality & Treatment.* 5th Ed. McGraw Hill: New York, NY

Bacchin P., M. Meireles, and P. Aimar. 2002. Modeling of Filtration: From the Polarised Layer to Deposit Formation and Compaction. *Desalination* 145:139-146.

Badruk M., N. Kabay, M. Demircioglu, H. Mordogan, and U. Ipekoglu. 1999a. Removal of Boron from Wastewater of Geothermal Power Plant by Selective Ion-Exchange Resins. I. Batch Sorption-Elution Studies. *Separation Science and* Technology 34(13):2553-2569.

Badruk M., Badruk M., N. Kabay, M. Demircioglu, H. Mordogan, and U. Ipekoglu. 1999b. Removal of Boron from Wastewater of Geothermal Power Plant by Selective Ion-Exchange Resins. II. Column Sorption-Elution Studies. *Separation Science and Technology* 34(15):2981-2995.

Baker J., T. Stephenson, S. Dard, and P. Cote. 1995. Characterization of Fouling of Nanofiltration Membranes Used to Treat Surface Waters. *Environmental Science and Technology*. 16: 977-985.

Batterman S., L. Zhang, and S. Wang. 2000. Quenching of Chlorination Disinfection By-Product Formation in Drinking Water by Hydrogen Peroxide. *Water Resources* 34(5):1652-1658.

Blau T., J.S. Taylor, K.E. Morris, and L.A. Mulford. 1992. DBP Control by Nanofiltration: Cost and Performance. *Journal AWWA* 84:104-116.

Bichsel Y. and U. Von Gunten. 1999. Oxidation of Iodide and Hypoiodous Acid in the Disinfection of Natural Waters. *Environmental Science and Technology* 33:4040-4045

Bichsel Y. and U. Von Gunten. 2000. Formation of Iodo-trihalomethanes during Disinfection and Oxidation of Iodide-Containing Waters. *Environmental Science and Technology* 34:2784-2791.

Bosh, A. 1998. Human Enteric Viruses in the Water Environment: a Minireview. *International Microbiology* 1:191-196.

Bowen W.R. and H. Mukhtar. 1996. Characterization and Prediction of Separation Performance of Nanofiltration Membranes. Journal of Membrane Science 112: 263-274.

Braghetta A., F.A. DiGiano, and W.P. Ball. 1998. NOM Accumulation at NF Membrane Surface: Impact of Chemistry and Shear. *Journal of Environmental Engineering ASCE* 124:1087-1098.

Brant J. and A. Childress. 2002. Assessing Short-range Membrane-colloid Interactions Using Surface Energetics. Journal of Membrane Science 203: 257-273.

Buecker B. 1997. *Power Plant Water Chemistry: A Practical Guide.* PennWell: Tulsa,OK.

Caballero S., F. X. Abad, F. Loisy, F. S. LeGuyader, J. Cohen, R. Pinto, and A. Bosh. 2004. Rotavirus Virus-Like Particles as Surrogates in Environmental Persistence and Inactivation Studies. *Applied and Environmental Microbiology* 70:3904-3909.

Callahan K.M., D.J. Taylor and M.D. Sobsey. 1995. Comparative Survival of Hepatitis A Virus, Poliovirus and Indicator Viruses in Geographically Diverse Seawaters. *Water Science and Technology* 31(5-6):189-193

Chae S. and C. Wang. 2002. Formation Characterization and Removal of Disinfection By-Products in Water Treatment Processes. *Water Science and Technology: Water Supply* 2(5):493-499.

Chang E.E., Y.P. Lin, and P.C. Chang. 2001. Effects of Bromide on the Formation of THMs and HAAs. *Chemosphere* 43:1029-1034.

Chellam S. 2000. Effects of Nanofiltration on Trihalomethane and Haloacetic Acid Precursor Removal and Speciation in Waters Containing Low Concentration of Bromide Ion. *Environmental Science and Technology* 34:1813-1820.

Chellam S and S.W. Krasner. 2001. Disinfection Byproduct Relationships and Speciation in Chlorinated Nanofiltered Waters. *Environmental Science and Technology* 35:3988-3999.

Childress A. and M. Elimelech. 1996. Effect of Solution Chemistry on the Surface Charge of Polymeric Reverse Osmosis and Nanofiltration Membranes. Journal of Membrane Science 199: 253-268.

Childress A. and M. Elimelech. 2000. Relating Nanofiltration Membrane Performance to Membrane Charge (Electrokinetic) Characteristics. Environmental Science and Technology 34: 3710-3716

Choi W.W. and K. Chen. 1979. Evaluation of Boron Removal by Adsorption on Solids. *American Chemical Society* 13(2): 189-196.

Corrosion Doctors. 1999. Langelier Saturation Index (LSI). Available <http://www.corrosion-doctors.org/NaturalWaters/Langelier.htm> [cited August 3, 2004]

Danavaro R. and M. Serresi. 2000. Viral Density and Virus-to-Bacterium Ratio in Deep-Sea Sediments of the Eastern Mediterranean. *Applied and Environmental Microbiology* 66:1857-1861.

Danavaro R., A. Dell'ano, A. Trucco, M. Serresi, and S. Vanucci. 2001. Determination of Virus Abundance in Marine Sediments. *Applied and Environmental Microbiology* 67:1384-1387.

Davies C.M., J. Long, M. Donald, and N.J. Ashbolt. 1995. Survival of Fecal Microorganisms in Marine and Freshwater Sediments. *Applied and Environmental Microbiology* 61:1888-1896.

Duckmanton L., S. Carman, E. Nagy, and M. Petric. 1998. Detection of Bovine Torovirus in Fecal Specimens of Calves with Diarrhea from Ontario Farms. *Journal of Clinical Microbiology* 36:1266-1270.

Espinasse B., P. Bacchin, and P. Aimar. 2002. On and Experimental Method to Measure Critical Flux in Ultrafiltration. *Desalination* 146:91-96.

Field R.W., D. Wu, J.A. Howell, and B.B. Gupta. 1995. Critical Flux Concept for Microfiltration Fouling. *Journal of Membrane Science* 100:259-272.

Fuhrman J. A. 1999. Marine Viruses and Their Biogeochemical and Ecological Effects. *Nature* 399:541-548.

Fuhrman J. A. 2001. Vector- & Water-Borne Disease. Oceans and Human Health Roundtable Report. Triangle Park, North Carolina: National Institute of Environmental Health Sciences and National Science Foundation.

Fuhrman J.A. and M. Schwalbach. 2003. Viral Influence on Aquatic Bacterial Communities. *Biology Bulletin* 204:192-195.

Gantzer C., E. Dubois, J.M. Crance, S. Billaudel, H. Kopecka, L. Schawartzbrod, M. Pommepuy, and F. LeGuyader. 1998. Influence of Environmental Factors on the Survival of Enteric Viruses in Seawater. *Oceanologica Acta* 21:983-993.

Garof H., R. Hewson, and D.J.E. Opstelten. 1998. Virus Maturation by Budding. *Microbiology and Molecular Biology Reviews* 62:1171-1190.

Garthwaite I. 2000. Keeping Shellfish Safe to Eat: A Brief Review of Shellfish Toxins, and Methods for their Detection. *Food Science and Technology* 11:235-244.

Griffin D. W., K.A. Donaldson, J.H. Paul, and J.B. Rose. 2003. Pathogenic Human Viruses in Coastal Waters. *Clinical Microbiology Reviews* 16:129-143.

Hara S., I. Koike, K. Terauchi, H. Kamiya, and E. Tanoue. 1996. Abundance of Viruses in Deep Oceanic Waters. *Marine Ecology Progress Series* 145:269-277.

Hoek E.M., M. Elimelech. 2003. Cake-Enhanced Concentration Polarization: A New Fouling Mechanism for Salt-Rejecting Membranes. *Environmental Science and Technology* 37:5581-5588.

Holmes-Farley R. 1998. Understanding Seawater. *Aquarium Frontiers.* July 1998

Howell J. 2004. Critical Flux Revisited. In *Fouling and Critical Flux*. Lappeenranta: Finland

Hughes K. A. 2003. Influence of Seasonal Environmental Variables on yhe Distribution of Presumptive Fecal Coliforms around an Antarctic Research Station. *Applied and Environmental Microbiology* 69:4884-4891.

Itoh M., S. Kunikane, and Y. Magara. 2001. Evaluation of Nanofiltration for Disinfection By-Products Control in Drinking Water Treatment. *Water Science and Technology: Water Supply* 1(5):233-243.

Jacangelo J.G., S.S. Adham, and J-M Laine. 1995. Mechanism of Cryptosporidium, Giardia, and MS-2 Virus Removal by MF and UF. *Journal AWWA* 87(9):107-121

Jiang S., R. Noble, and W. Chu. 2001. Human Adenoviruses and Coliphages in Urban Runoff-Impacted Coastal Waters of Southern California. *Applied and Environmental Microbiology* 67:179-184.

Kitis M., J.C. Lozier, J-H Kim, B. Mi, and B.J. Mariñas. 2003. Microbial Removal and Integrity Monitoring of RO and NF Membranes. *Journal AWWA* 95(12):105-119

Kampioti A.A. and E.G. Stephanou 2002. The Impact of Bromide on the Formation of Neutral and Acidic Disinfection By-Products (DBPs) in Mediterranean Chlorinated Drinking Water. *Water Research* 36:2596-2606.

Kedem, O. and A. Katchalsky. 1958. Thermodynamic Analysis of the Permeability of Biological Membranes to Non-Electrolytes. *Biochim. Biophys. Acta* 27:229-246.

Kwok D.Y., T. Gietzelt, K. Grundke, H.J. Jacobasch, and A.W. Neumann. 1997. Contact Angle Measurements and Contact Angle Interpretation. 1. Contact Angle Measurements by Axisymmetric Drop Shape Analysis and a Goniometer Sessile Drop Technique. Langmuir 13: 2880-2894.

Lane R.W. 1993. *Control of Scale and Corrosion in Building Water Systems.* McGraw-Hill, Inc.: New York, NY

Lee S., A. Gary, and J. Cho. 2004. Applicability of Sherwood correlations for natural organic matter (NOM) transport in nanofiltration (NF) membranes. *Journal of Membrane Science* 240:49-65.

Lee S., J. Cho, and M. Elimelech. 2004b. Influence of Colloidal Fouling and Feed Water Recovery on Salt Rejection of RO and NF Membranes. *Desalination* 160:1-12.

Li H., A.G. Fane, H.G.L. Coster, and S. Vigneswaran. 2000. An Assessment of Depolarisation Models of Crossflow Microfiltration by Direct Observation Through the Membrane. *Journal of Membrane Science* 172:135-148.

Magara Y., A. Tabata, M. Kohki, M.. Kawasaki, and M. Hirose. 1998. Development of Boron Reduction System for Sea Water Desalination. *Desalination* 118:25-34.

Mallevialle J., P.E. Odendall, and M.R. Wiesner. 1996. *Water Treatment Membrane Processes.* McGraw-Hill: New York, NY.

Manttari M. and M. Nystrom. 2000. Critical Flux in NF of High Molar Mass Polysaccharides and Effluents from the Paper Industry. *Journal of Membrane Science* 170:257-273.

Marangou V.S. and K. Savvides. 2001. First Desalination Plant in Cyprus - Product Water Aggressivity and Corrosion Control. *Desalination* 138:251.

Matsammuuronen S., J.A. Howell, and M. Nyström. 2002. Critical Flux in Ultrafiltration of Myoglobin and Bakers Yeast. *Journal of Membrane Science* 196:13-26.

Maugeri T.L., M. Carbone, M.T. Fera, G.P. Irrera, and C. Gugliandolo. 2004. Distribution of Potentially Pathogenic Bacteria as Free Living and Plankton Associated in a Marine Coastal Zone. *Journal of Applied Microbiology* 97:354-361.

Meijers R.T. 1995. Potential Treatment Options for Restriction of Bromate Formation and Bromate Removal. *Water Supply* 13(1):183-189.

Melnik L., O. Vysotskaja, and B. Kornilovich. 1999. Boron Behavior During Desalination of Sea and Underground Water by Electrodialysis. *Desalination* 124:125-130.

Mos L. 2001. Domoic Acid: A Fascinating Marine Toxin. *Environmental Toxicology and Pharmacology* 9:79-85.

Moss P. and J.M. Gil. 1999. Twenty Five Years of Desalination in The Canary Islands: A Historical Review of the Application of Reverse Osmosis Using Case Studies and Operation Experience. *Desalination* 125:17-23.

Mulder M. 1996. *Basic Principles of Membrane Technology.* Kluwer Academic Publishers: Boston, MA.

Nadav N. 1999. Boron Removal from Seawater Reverse Osmosis Permeate Utilizing Selective Ion Exchange Resin. *Desalination* 124:131-135.

Ning R.Y. and J.P. Netwig. 2002. Complete Elimination of Acid Injection in Reverse Osmosis Plants. *Desalination* 143:29.

Noble R. and J.A. Fuhrman. 1997. Virus Decay and its Causes in Coastal Waters. *Applied and Environmental Microbiology* 63:77-83.

Noble R., and Fuhrman, J. A. 2000. Rapid Virus Production and Removal as Measured with Fluorescently Labeled Viruses as Tracers. *Applied and Environmental Microbiology* 66:3790-3797

Pastor M.R., A.F. Ruiz, M.F. Chillon, and D.Prats. 2001. Influence of pH in the Elimination of Boron by Means of Reverse Osmosis. *Desalination* 140:145-152.

Paytan A. 2004. Marine Chemistry Lecture 5: Trace Elements in Seawater. Available <http://ocean.stanford.edu/bomc/info.html> [cited July 12, 2004].

Plottu-Pecheux A., C. Democrate, B. Houssais, D. Gatel, and J. Cavard. 2001. Controlling the Corrosiveness of Blended Waters. *Desalination* 138:237.

Praptowidodo V.S. and A. Khalik. 2000. Studies on Seawater Desalination by Reverse Osmosis at the Badak Natural Gas Liquefaction Plant, Bontang, East Kalimantan. *Desalination* 132:323.

Prats D., M.F. Chillon-Arias, M. Rodriguez-Pastor. 2000. Analysis of the Influence of pH and Pressure on the Elimination of Boron in Reverse Osmosis. *Desalination* 128:269-273.

Procotor L.M. and J.A. Fuhrman. 1991. Roles of Viral Infection in Organic Particle Flux. *Marine Ecology Progress Series* 69:133-142.

Puckorius P.R. and J.M. Brooke. 1991. A New Practical Index for Calcium Carbonate Scale Prediction in Cooling Tower Systems. *Corrosion* 47:280.

Sablani S.S, M.F.A. Goosen, R. Al-Belushi, and M. Wilf. 2001. Concentration Polarization in Ultrafiltration and Reverse Osmosis: A Critical Review. *Desalination* 141: 269-289.

Schafer A., 2001. *Natural Organics Removal Using Membranes.* Techonomic: Lancaster, Pennsylvania.

Seidel Arza and M. Elimelech. 2002, Coupling Between Chemical and Physical Interaction in Natural Organic Matter (NOM) Fouling of Nanofiltration Membranes; Implications for Fouling Control. *Journal of Membrane Science* 203:245-255.

Shukairy H.M, R.J. Miltner, and R.S. Summers. 1994. Bromide's Effect on DBP Formation, Speciation, and Control: Part 1, Ozonation. *Journal AWWA* 86:72-87.

Simonnot M.O., C. Castel, M. Nicolai, C. Rosin, M. Sardin, and H. Jauffret. 2000. Boron Removal from Drinking Water with a Boron Selective Resin: Is the Treatment Really Selective? *Water Resources* 34(1):109-116.

Singer C. 1994. Control of Disinfection By-Products in Drinking Water. *Journal of Environmental Engineering* 120(4):727-744.

Sinton L.W., R.K. Finlay, and P.A. Lynch. 1999. Sunlight Inactivation of Fecal Bacteriophages and Bacteria in Sewage-Polluted Seawater. *Applied and Environmental Microbiology* 65:3605-3613

Sketchell J., H.G. Peterson, and N. Christofi. 1995. Disinfection By-Product Formation After Biologically Assisted GAC Treatment of Water Supplies with Different Bromide and DOC Content. *Water Research* 29(12):2635-2642.

Sobsey M.D., K.J. Schwab, and T.R. Handzel. 1990. A Simple Membrane Filter Method to Concentrate and Enumerate Male-Specific RNA Coliphages. *Journal AWWA* 82(9):52-59

Taniguchi M., Y. Fusaoka, T. Nishikawa, and M. Kurihara. 2004. Boron removal in RO seawater desalination. *Desalination* 167: 419-426.

Trofe T.W., G.W. Inman, and J.D. Johnson. 1980. Kinetics of Monochloramine Decomposition in The Presence of Bromide. *Environmental Science and Technology* 14(5):544-549

Tseng T., R.C. Cheng, D.X. Vuong, and K.L. Wattier. 2005. Developing an Experimental Protocol for Evaluating Low-Pressure Desalting Membranes for Seawater Desalination. In *Proc. AWWA Membrane Technology Conference*, Phoenix, AZ.

M. Turan. 2004. Influence of Filtration Conditions on the Performance of Nanofiltration and Reverse Osmosis Membranes in Dairy Wastewater Treatment. Desalination 170: 83-90.

U.S. Bureau of Reclamation. 2003. *Desalting Handbook for Planners. 3rd Edition*. Denver, CO: U.S. Dept of Interior, Bureau of Reclamation, Water Treatment Engineering and Research Group

U.S. Environmental Protection Agency. 2001. Method 1601: Male-Specific (F +) and Somatic Coliphage in Water by Two-Step Enrichment Procedure. Available <http://www.epa.gov/microbes/1601ap01.pdf> [cited April 27, 2005]

U.S. Environmental Protection Agency. 1991. Guidance Manual for Compliance with the Filtration and Disinfection Requirement for Public Water Systems Using Surface Water Sources.

Vrijenhoek E.M., S. Hong, and M. Elimelech. 2001. Influence of Membrane Surface Properties on Initial Rate of Colloidal Fouling of Reverse Osmosis and Nanofiltration Membranes. *Journal of Membrane Science* 188:115-128.

White G.C. 1986. *Handbook of Chlorination* 2nd Ed. Van Nostrand Reinhold Company: New York, NY

Wu D., J.A. Howell, R.W. Field. 1999. Critical Flux Measurement for Model Colloids. *Journal of Membrane Science* 152:89-98.

Yang G., W. Xing, and N. Xu. 2003. Concentration Polarization in Spiral-Wound Nanofiltration Membrane Elements. *Desalination* 154:89-99.

You S.-H., D.-H. Tseng, and G.-L. Guo. 2001. A Case Study on the Wastewater Reclamation and Reuse in the Semiconductor Industry. *Resources, Conservation, and Recycling* 32:73.

Zingone A. and H. Oksfeldt. 2000. The Diversity of Harmful Algal Blooms: A Challenge for Science and Management. *Ocean and Coastal Management* 43:725-748.

ABBREVIATIONS

A	membrane (surface)
AAS	atomic absorption spectrometry
AFM	atomic force microscopy
ATCC	american type culture collection
ASP	amnesic shellfish poisoning
Alk	Alkalinity
amu	atomic mass unit
AWWA	American Water Works Association
AwwaRF	Awwa Research Foundation
C	Concentration
°C	degrees Celsius
CCPP	Calcium Carbonate Precipitation Potential
CDHS	California Department of Health Services
CD	compact disc
CD-ROM	compact disc-read-only memory
cfu	colony forming unit
Co.	Company
CP	concentration polarization
CPM	concentration polarization modulus
CRW	Colorado River Water
C•T	disinfectant concentration times residence time
DBP	disinfection byproduct
DDW	doubly deionized water
DI	deionized water
DLS	dynamic light scattering
DW	desalinated water
°F	degrees Fahrenheit
ft	Foot
ft^2	square foot
ft^3	cubic foot
ft^3/min	cubic foot per minute
ft/s	foot per second
g	Gram
gal	Gallon
gfd	gallon per square foot per day
g/L	gram per liter
GPC	gel permeation chromatography
gpd	gallon per day
gpm	gallon per minute

gpm/ft^2	gallon per minute per square foot
GW	Groundwater
HAA	haloacetic acid
HAA5	five haloacetic acids
hp	horsepower
hr	hour
IC	ion chromatography
ICP	inductively coupled plasma
IDBP	iodinated disinfection byproducts
in.	inch
in.^{2}	square inch
IP	ion product
I-THM	iodinated trihalomethane
IX	ion exchange
J	permeate flux
J_w	water flux
J_i	flux of salt *i*
Jour.	Journal
k	mass transfer coefficient
KGal	1,000 gallons
kWh	kilo-watt hours
K_{sp}	solubility product
L	liter
LBWD	Long Beach Water Department
L/h	liter per hour
L/min	liter per minute
L/s	liter per second
LSI	Langelier Saturation Index
m	meter
m^2	square meter
m^3	cubic meter
M	molar
m^3/d	cubic meter per day
MF	microfiltration
m/h	meter per hour
m^3/h	cubic meter per hour
mg	milligram
mgd	million gallons per day
mg/L	milligram per liter
ML	megaliter

ML/d	megaliter per day
mL	milliliter
mL/s	milliliter per second
mm	millimeter
mm^2	square millimeter
$m^3/m^2/h$	cubic meter per square meter per hour
MS	male specific
m/s	meter per second
mm/s	millimeter per second
m^3/s	cubic meter per second
mV	millivolt
MW	molecular weight
MWCO	molecular weight cutoff
μm	micrometer
n	number of dissociating ions
NACE	National Association of Corrosion Engineers
ND	not detected
NF	nanofiltration
NF^2	dual-staged nanofiltration
NOM	natural organic matter
ntu	nephelometric turbidity unit
P	pressure
p.	page
PAC	Project Advisory Committee
P.E.	Professional Engineer
Ph.D.	Doctor of Philosophy
pp.	pages
ppm	part per million
Proc.	Proceedings
psi	pound per square inch
π	osmotic pressure
Q	flow
R	rejection
R_a	average roughness
R_f	resistance due to concentration polarization and fouling effects
R_g	gas constant
R_{ir}	irreversible fouling resistance
R_m	clean membrane resistance
RO	reverse osmosis
R_q	root-mean-squared roughness
R_r	reversible fouling resistance
RSI	Ryznar Saturation Index

RMS	root-mean-squared
RTW	Rothberg, Tamburini and Windsor
SAD	surface area difference
SEC	size exclusion chromatography
SEM	scanning electron microscopy
SI	Système International [International System of Units]
spp.	species
SPW	State Project Water
SW	surface water
SWNF	seawater nanofiltration
SWRO	seawater reverse osmosis
SWTR	Surface Water Treatment Rule
σ	membrane reflection coefficient
T	temperature
TDS	total dissolved solids
THM	trihalomethane
TMP	transmembrane pressure
TOC	total organic carbon
TTHM	total trihalomethanes
UF	ultrafiltration
UNR	University of Nevada, Reno
U.S.	United States
U.S.A.	United States of America
USEPA	United States Environmental Protection Agency
Y	recovery

IWA Publishing's authorised EU representative for General Product Safety Regulations is Diane D'Arras, 15 rue Duret, 75116 Paris, France, e-mail: safety@iwap.co.uk.

Printed and bound by CPI Group (UK) Ltd, Croydon, CR0 4YY
17/09/2026
02218798-0016